LYN WHITAKER holds a B.Sc. degree in botany from the University of Manchester and was an examiner and teacher of biology at Rickmansworth Grammar School for several years.

JANET KELLY is head of the Biology Department at Rickmansworth School, which she joined after gaining a B.Sc. degree in zoology at the University of Nottingham. An examiner in CSE and GCE O-level human biology and CSE environmental studies, she is the author of the *General Science* cards and the *Human Biology Course Companion* in the Key Facts series.

D0494016

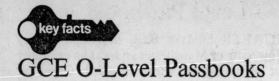

GCE O-Level Passbooks

CHEMISTRY, C. W. Lapham, M.Sc., A.R.I.C.

MODERN MATHEMATICS, A. J. Sly, B.A.

HISTORY (*Social and Economic*, 1815–1939), M. C. James, B.A.

FRENCH, G. Butler, B.A.

ENGLISH LANGUAGE, Robert L. Wilson, M.A.

GEOGRAPHY, R. Knowles, M.A.

PHYSICS, B. P. Brindle, B.Sc.

GCE O-Level Passbook

Biology

R. Whitaker, B.Sc. and
J. M. Kelly, B.Sc., M.I. Biol.

Published by Intercontinental Book Productions
in conjunction with Seymour Press Ltd.

Distributed by Seymour Press Ltd.,
334 Brixton Road, London, SW9 7AG

This book is sold subject to the condition that it shall not, by way of trade or otherwise, be lent, re-sold, hired out, or otherwise circulated without the publisher's prior consent in any form of binding or cover other than that in which it is published and without a similar condition including this condition being imposed on the subsequent purchaser

Published 1976 by Intercontinental Book Productions, Berkshire House, Queen Street, Maidenhead, Berks., SL6 1NF in conjunction with Seymour Press Ltd.

1st edition, 1st impression 2.76.0
Copyright © 1976 Intercontinental Book Productions
Made and printed by C. Nicholls & Company Ltd
ISBN 0 85047 906 1

Contents

Contents

Introduction

Biology is the study of living things or organisms. For many years it has been divided into zoology, the study of animals, and botany, the study of plants. In more recent years a third division, microbiology, has been included. This is the study of microscopic organisms such as bacteria and viruses. The characteristics by which living things are recognised are nutrition, respiration, excretion, movement, sensitivity, growth and reproduction. These in fact constitute the framework of the GCE O-level and CSE syllabuses. A knowledge of these characteristics in a mammal and a flowering plant is required, along with the study of a number of selected organisms. In this way the most highly specialised representatives of the plant and animal kingdoms (the mammals and flowering plants) are studied in some depth, while a more superficial study is made of the rest of the plant and animal kingdoms. A similar framework has been adopted for this book. The authors have tried, wherever possible, to relate structure to function, and hope that the material builds up an overall picture of how organisms survive.

However, in the last fifty years biology has undergone some very important changes. These have resulted not only in the establishment of new facts, but also in the development of a new approach to the subject. As a result of the advent of the electron microscope, and of improved biochemical techniques such as paper chromatography and the use of radioactive isotopes, biologists have been able to study the functioning of organisms at the cellular and even molecular level. This has revealed a uniformity within plant and animal cells, and has focused the biologist's attention on the cell, rather than on the entire organism, as the unit of life. For this reason the first chapter deals with the area where exciting developments have recently occurred. The existence of the gene was established at the beginning of this century, but the method by which it carries out its action is only gradually being understood. A breakthrough in the study of the gene came in the 1950s when Crick and Watson, working at Cambridge, established the structure of the DNA molecule.

The last chapter, on ecology, is concerned with the relationships of organisms with their environment. Some topics which are of great concern to mankind at present are raised briefly here. Man is gradually realising the problems caused by certain aspects of his way of life, and the danger threatening both the human and other populations. The energy crisis, pollution and starvation are among the major causes of concern. Biologists have a very important part to play in making the world aware of them.

Biology is an essential part of the training of doctors, veterinary surgeons and agriculturalists, among others. In addition, since it helps in the understanding of the human body and of the relationships between man and his environment, it is a subject that many students find increasingly relevant today. Newspaper articles on organ transplants or on the possible extinction of some species of wild life are now fairly commonplace: with some basic biological knowledge such topics will be much more meaningful.

Units used in this book

S.I. (International System) units are used throughout. The abbreviations are:

m = metre
mm = millimetre (0·001 m)
μm = micrometre (0·000001 m) = one old micron unit (μ)
nm = nanometre (0·000000001 m)
g = gram
mg = milligram
J = joule
mm^3 = cubic millimetre

Chapter 1
The Cell and Cell Organisation

The word 'cell' was first used by Robert Hooke in 1665. He built one of the earliest compound microscopes, which enabled him to see box-like arrangements inside thin slices of substances such as cork. He called these small boxes cells. This discovery illustrates well the fact that advances in biological science are often dependent on progress in other sciences. Biologists need tools and techniques developed by physicists or chemists, such as the electron microscope and radioactive isotopes. Biology is not an isolated science, but one with links with other sciences, including mathematics and chemistry, and this is becoming increasingly true in modern biology.

The observations made by Hooke were extended, and it became apparent that most living things were composed of cells. In 1839 a theory was proposed stating that cells were the basic units of life. There are of course a number of exceptions to this generalisation, but these are very much in the minority. Under the light microscope a typical plant and animal cell look as in figure 1.

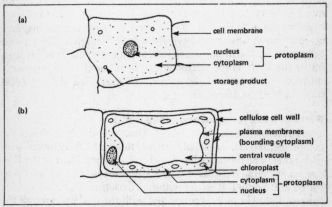

Figure 1. Animal (a) and plant (b) cells (magnification about × 400)

In all cells the cytoplasm is a jelly-like substance containing many particles. It can change its consistency, and in some cases appears to flow around the cell. The processes of life occur within the cytoplasm. The nucleus is embedded in the cytoplasm and is usually ovoid in shape. It controls the development and activity of the cell, and no cell can exist for long without its nucleus. The plasma membrane is the living boundary in all cells, and its structure and function are discussed later. The vacuole in plant cells is non-living. It is a fluid-filled space which pushes the cytoplasm out against the cell wall. This helps in giving support to plant tissues. The fluid in the vacuole consists of salts, sugars, etc. dissolved in water. The average size of an animal cell is around 0·01 mm in diameter. There are larger cells, just visible to the human eye, of about 0·1 mm in diameter, and cells which are smaller, of about 0·001 mm in diameter. A light microscope can magnify these cells by up to 1,500 times.

As can be seen from figure 1 there are a number of basic differences between the typical plant and animal cells. Plant cells tend to be larger than animal cells, and more clearly defined, owing to the presence of a dead cellulose cell wall. The centre of a plant cell consists of a vacuole, but only temporary vacuoles occur in animal cells. Many plant cells contain chloroplasts and starch grains, but these never occur in animal cells.

The study of cells (**cytology**) was greatly advanced by the development of a completely new type of microscope, the electron microscope (E.M.). This was not in widespread use in research departments until the late 1950s. These microscopes can magnify up to 500,000 times, and have altered the biologists' ideas about cell structure. It is important to look at photographs of cells as seen under the E.M. There are many excellent books containing such micrographs. A generalised animal cell magnified about 10,000 times by the E.M. looks as in figure 2.

Parts of a cell (as seen under the E.M.)
Endoplasmic reticulum (E.R.) This extends throughout the cytoplasm, and is made up of flattened tubes which are linked with each other. The tubes are lined with membranes. Most of the E.R. has on its surface particles called **ribosomes**, and is called the rough E.R. Some E.R. exists without ribosomes on it and is called smooth E.R. The E.R. is connected with the nucleus through the gaps and pores in the nuclear membrane, supporting the idea that the function of the E.R. is a kind of transport system within the cells. Substances can move between the nucleus and cytoplasm, and throughout the cytoplasm within the E.R. This freedom of movement is important in the synthesis and transport of proteins.

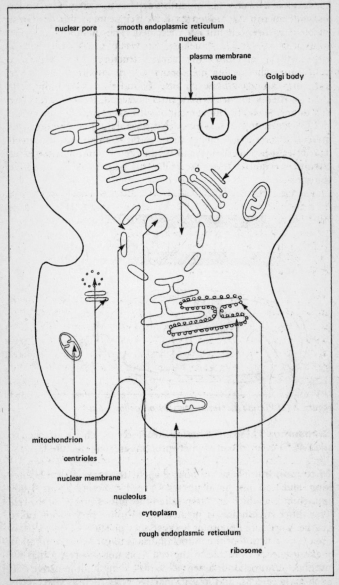

Figure 2. Electron micrograph of an animal cell (magnification about × 10,000)   11

Plasma membrane Although this appears as a single line under the light microscope, under the E.M. it appears as three regions, which have been chemically identified. The membranes which occur around the E.R., nucleus, sap vacuole, chloroplasts, and mitochondria all have the same structure as the plasma membrane. The plasma membrane is very important since it is the living boundary between all cells and their surroundings. It controls chemically what may enter and leave the cell. How the membrane allows the movement of substances into and out of the cell is not certain, but it is likely that several different methods are involved. Certain substances seem to be pumped out of the cell and other substances appear to be dragged in, and in these cases energy is required from the cell. This process is called active transport.

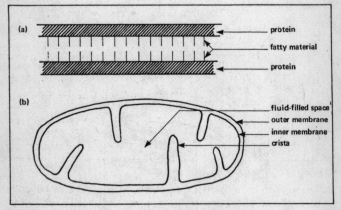

Figure 3. (a) Plasma membrane and (b) mitochondrion

Ribosomes These are spherical structures which are attached to the E.R. They are the sites where proteins are manufactured.

Mitochondria These are boat-shaped structures about $2\cdot5$ μm long and $1\cdot0$ μm in diameter. A cell contains about 1,000 mitochondria which are often called the 'power house' of the cell. The inner membrane is pushed in, forming projections called cristae. These are important because they produce a larger surface area. The function of the mitochondria is to carry out respiration, making energy available to the cell. It is not surprising that the number of mitochondria per cell varies, being high in active cells such as liver cells, and low in inactive cells such as cheek cells.

The enzymes needed for some of the stages in respiration are situated in the mitochondria, being attached to the inner membrane. The longer the inner membrane, the more enzymes can be packed in.

Golgi body (apparatus, or complex) These are areas within the cytoplasm which are detectable even with the light microscope. They consist of piles of flattened spaces lined with smooth E.R. and also many smaller rounded droplets. These areas contain materials produced by the cell, and it is thought that these materials are carried out of the cell by the Golgi body.

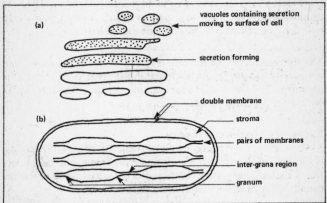

Figure 4. (a) Golgi body and (b) chloroplast

Chloroplasts These are relatively large structures, easily visible with the light microscope in most plant cells. They are never present in animal cells. They are about 5 μm in diameter. Each chloroplast contains about 60 grana. The green pigment, chlorophyll, is contained between the lamellae. All the enzymes needed for the complex reactions of photosynthesis are present in the chloroplasts. Those involved in the dark reaction are localised in the stroma, those needed for the light reaction in the lamellae. The end products of photosynthesis accumulate in the stroma.

Cell wall The plant cell wall is a non-living structure whose primary function is that of support. A cell wall is never present in animal cells. The first sign of the development of a cell wall, after cell division, is a thin layer called the middle lamella. On this layer cellulose is deposited inwards. Cellulose is a complex carbohydrate made up of strands which are built up in layers to form the cell wall. The wall may have extra substances added, such as the impermeable waxy cuticle in epidermal cells.

The nucleus is separated from the cytoplasm by the nuclear membrane. The nucleus contains the chromosomes which are the 'blue-print' for the cell's activities. This blue-print is known as the genetic material. The nucleolus is also present in the nucleus.

Centrioles exist in pairs at right angles to each other. They are cylindrical structures and take part in cell division in animals. In non-dividing cells they lie just outside the nucleus.

The micro-structure of the cell, revealed by the E.M., shows that division of labour occurs within the cell, as it does within the organism. The chloroplasts, for example, are specialised structures which carry out only photosynthesis.

Exceptions to the cell theory
A cell may conveniently be defined as a nucleus and its surrounding cytoplasm enclosed within a plasma membrane. The cell theory states that organisms are composed of cells, and these in fact are usually specialised for different functions.

There are however a few exceptions to this theory. The unicellular organisms such as *Amoeba* or *Chlorella* cannot be considered as being divided up into cells, and all the living processes occur within the single unit. Frequently in these organisms different regions of the protoplasm have become specialised for different functions. Some organisms, such as fungi, consist of a mass of cytoplasm in which many nuclei are embedded, with no partitions separating off units: these are called coenocytes. Bacteria are unusual in that not only are they unicellular but they contain no clearly defined nucleus. The genetic material exists as strands lying free in the cytoplasm and not enclosed by a nuclear membrane. Viruses, which are thought of as being on the boundary between living and non-living things, are unicellular and also structurally and chemically unlike the typical cell. Materials such as cork are rather an oddity, for although made up of cells, the cells are dead: i.e. not all things which are made up of cells are, in fact, living things.

Tissues

In multicellular organisms cells do not exist in isolation but are grouped together to form tissues. **A tissue** is a collection of similar cells performing similar functions.

Mammalian tissues

Within the body of a mammal the main types of tissue which can be recognised are epithelial, skeletal, muscular and nervous. Blood and lymph are fluid tissues and will be discussed in chapter 4. It is important to remember how the structure of a tissue is related to its function.

Epithelial tissue lines or covers; it consists of prominent cells lying on a basement membrane, and two basic types can be recognised: **simple** epithelium consisting of one layer of cells, and **compound** epithelium consisting of more than one layer of cells. Simple epithelium can be further subdivided into pavement (squamous), cubical, columnar and ciliated.

Squamous epithelium consists of flat cells which give a crazy paving effect from above; it is found where substances are required to diffuse across it and the thin cells ensure that the diffusion distances are small and diffusion is therefore rapid. It occurs in the alveoli and capillary walls across which the exchange of respiratory gases occurs (see pages 35 and 94), and in the kidney. In **cubical epithelium** the cells are square in outline; many glands and their ducts are lined by it and it is also found in the kidney tubules. **Columnar epithelium** consists of long column-like cells and it lines the intestine. In the ileum the free edge of the cells has many folds called **microvilli** which increase the surface area for the absorption of digested foods. In **ciliated epithelium** the cells are columnar in shape but the free edge bears numerous hair-like structures called **cilia** which are capable of beating rhythmically. This type of epithelium is found where substances have to be moved; for example, cilia in the respiratory tract move mucus, which has trapped dust and micro-organisms from the inhaled air, away from the lungs. Epithelia can also be modified to form **glandular tissue**; for example, the goblet or mucus-producing cells which are found among the columnar epithelium of the intestine and the ciliated epithelium of the respiratory tract.

Stratified epithelium is a compound epithelium and it

comprises the epidermis of the skin. The epidermis consists of three cell layers and the lower **Malpighian layer** consists of actively dividing cells; the newly formed cells move outward and become part of the living **granular layer**, the upper cells of which become incorporated into the **cornified layer** as they become impregnated with **keratin** and die; this substance is tough and impermeable and enables the skin to act as an efficient protective barrier. Another type of compound epithelium is **transitional epithelium**; it is found in the bladder and allows it to increase in size as it fills up with urine.

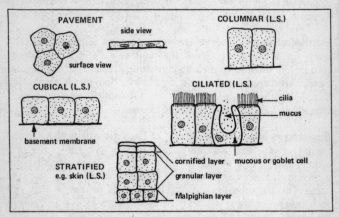

Figure 5. The structure of epithelial tissue

Two types of **skeletal tissue** occur in vertebrates: bone and cartilage. **Cartilage** forms the permanent skeleton in sharks but in mammals the skeleton is composed largely of bone with cartilage persisting in places, for example at the ends of bones. Cartilage (gristle) is a glassy substance consisting of a ground substance (matrix) in which are embedded cartilage cells lying singly or in groups of two or four. Each cell lies in a space or lacuna.

Bone is much harder than cartilage and this is attributed to the calcium salts, mainly **calcium phosphate**, which impregnate the matrix. The salts are secreted by bone cells (osteoblasts) which in compact bone lie in spaces and are arranged in concentric rings around a central Haversian canal which contains blood vessels. Such a structure is called a **Haversian system** and there are many of them arranged side by side.

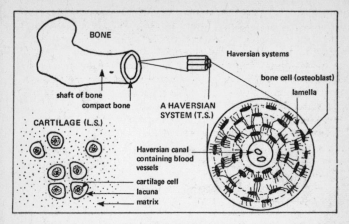

Figure 6. The structure of bone and cartilage

Muscular tissue contains contractile fibrils and is adapted to contract and relax; there are three types of muscle: unstriated, striated and cardiac. **Unstriated muscle** (also known as unstriped, visceral or smooth muscle) consists of densely packed, elongated cells or fibres each with a single nucleus and containing many fine contractile fibrils. It does not have a striped appearance. It is found in the walls of the gut and blood vessels; it contracts relatively slowly, is not easily fatigued and is not under conscious control. **Striated muscle** (also known as striped or skeletal muscle) consists of very elongated fibres each with many nuclei; the arrangement of the numerous contractile fibrils gives the fibres a striped appearance. Striped muscle is associated with the skeleton and is important in locomotion. It is stimulated by nerves, contracts quickly, easily fatigues and is under conscious control. **Cardiac muscle** is found in the heart; it consists of fibres which are similar in appearance to those of striated muscle but they are joined to each other by cross connections.

Nervous tissue is adapted to conduct impulses (messages) and it consists of **neurones** or nerve cells. Neurones differ in structure according to their position in the central nervous system (C.N.S.) but certain common features can be recognised. Each has a cell body from which processes extend; the cell body consists of a nucleus and cytoplasm. Branching processes called **dendrons** conduct impulses towards the cell body and **axons** conduct impulses away from the cell body. The axons can be several feet in

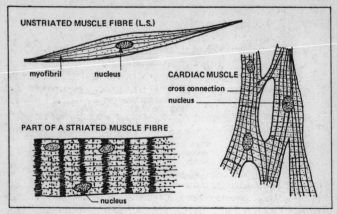

Figure 7 (a). The structure of muscle

length and may be enclosed in a fatty myelin sheath which insulates them and speeds up the rate of conduction. A **sensory neurone** conducts impulses from sensory receptors towards the C.N.S. and the cell body lies in a ganglion outside the C.N.S. **Motor neurones** conduct impulses from the C.N.S. to effector organs (muscle or gland) and the cell body is in the C.N.S. **Intermediate (connector) neurones** connect sensory and motor neurones and are mainly confined to the C.N.S. A **synapse** is where one neurone meets another.

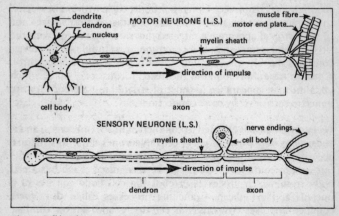

Figure 7 (b). The structure of nervous tissue

18

Plant tissues

There is a smaller variety of tissues in plants than in animals. This is related to their different methods of feeding. Animals in general have more complex methods of detecting and then moving rapidly towards their food. Plant tissues do not fall neatly into clearly defined types, but for convenience they will be dealt with under the headings of dividing, packing, transporting, supporting and protecting tissues. These tissues have technical names; it is not important to remember the names but to know how the structure of the tissue performs its function well (i.e. how structure is related to function).

1. Meristematic tissues (dividing cells) These cells usually occur at the tips of roots and shoots, and are responsible for enabling these organs to increase in length. Dividing cells also make up the **cambium**, and are responsible for the increase in the girth of roots and shoots. They are small, thin-walled cells with no central vacuole or chloroplasts. They retain their ability to divide and so allow growth to occur.

2. Parenchyma (packing cells) These cells literally fill the spaces between other tissues. When the cells are fully expanded with water (turgid) they also add to the firmness of the organ. Some organs depend almost entirely on this method of support. The fact that a cut bluebell flower stalk droops or wilts when deprived of water is a good example of this. Parenchyma cells are rounded and lightly packed together with air spaces between them. They may contain storage products such as starch. They have cellulose cell walls, are living cells, and are not elongated in any direction. Photosynthetic tissue is sometimes listed as a separate tissue, but in fact is just a collection of parenchyma cells which contain chloroplasts.

3. Vascular tissue (transporting cells) There are two types of transporting cells in plants which carry out the equivalent function of the circulatory system in animals. **Xylem tissue** is made up mainly of elongated tubes, called xylem vessels. These are dead structures because an extra strengthening structure called **lignin** has been added to the cellulose cell wall. This makes the tubes waterproof and kills the living protoplasm. Some gaps, called **pits**, may remain in the lignin. These pits allow water and salts to move out of the xylem vessel into the adjacent tissues. Water and mineral salts are transported upwards in the xylem vessels, from the roots to the leaves. There are no horizontal cross walls in these vessels to slow down the progress of water movement. **Phloem tissue** is made up mainly of living cells called sieve tubes. Sieve tubes, like xylem vessels, develop from columns of cells, but the horizontal cross walls do not break down completely, they only become perforated (hence the name sieve tubes). These sieve areas allow soluble food materials to move from the leaves to all other parts of the plant. This may involve upward or downward movement. Xylem and phloem tissues also include some packing and supporting tissues, and are often called veins. Phloem sieve tubes have cellulose cell walls, and very active cells called companion cells closely linked with them. It is thought that the companion cell must carry out the living processes for the sieve tube, in particular providing it with energy for translocation.

4. Supporting tissue The xylem vessels, because they have lignified walls, play an important role in support. In addition, many plants contain fibres, mainly in their stems. These are very elongated structures, with thick lignified walls which give them strength. The ends of the fibres are tapered and overlap with other fibres. Their technical name is **sclerenchyma**. Another type of supporting tissue, which is more flexible than fibres, is called **collenchyma**. These are living cells, with extra bands of cellulose thickening, usually at the corners of the cell wall. This tissue is positioned near the surface of a stem, where more bending is experienced.

5. Protecting tissue Young stems and leaves are protected on their surface by **epidermal cells**. These are flattened cells with no gaps between them. Their outer cellulose cell walls have a waxy cuticle covering them, which reduces water loss. The epidermis also protects against infection from micro-organisms. Older stems, as they increase in girth, rupture the epidermis and its protective function is replaced by the **bark**. The bark is a dead corky layer which is constantly being added to by the cork cambium. Since the epidermis and bark are impermeable not only to water, but also to gases, they must have gaps in them which allow for gas exchange. In the case of the epidermis, these are the **stomata** (singular **stoma**), and in bark air spaces called **lenticels** are present.

On the following two pages diagrams are given of the main plant tissues. Transverse sections (T.S.) and longitudinal sections (L.S.) are shown. Transverse sections are made at right angles to the long axis of the organ, and longitudinal sections are made parallel to the long axis of the organ.

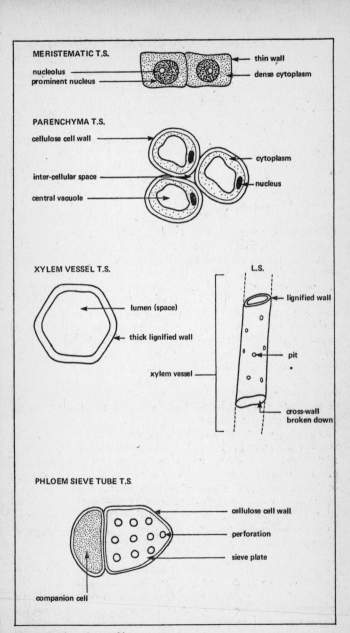

Figure 8. Plant tissues (I)   21

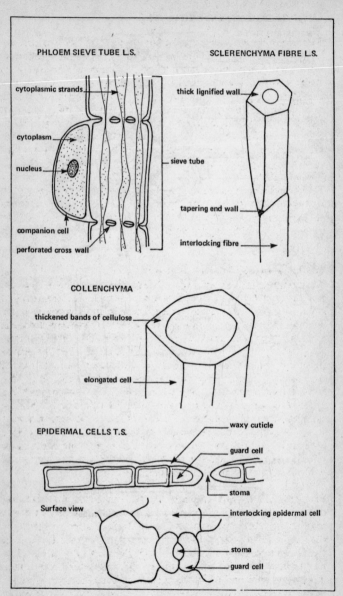

Figure 9. Plant tissues (II)

22

Organs and organ systems

Tissues rarely work in isolation but are arranged into organs. An **organ** is a distinct structure formed from a collection of different tissues and it has one or more specific functions to perform. Each tissue makes its own particular contribution to the functioning of the organ as a whole. In mammals, for example, the stomach is an organ; it contains epithelial tissue, unstriated muscle, nervous tissue and blood, all organised into a complex structure which has the function of food storage and protein digestion. Fewer organs are recognised in plants; the leaf is an organ, the function of which is to carry out photosynthesis.

In the body organs work in conjunction with each other as part of an **organ system** which will have one or more specific functions to perform. For example, the digestive system consists of the alimentary canal and other organs such as the pancreas and liver, and its function is to digest and absorb food.

Levels of organisation in animals

Most organisms have achieved the organ level of organisation and possess organ systems, but other levels do exist. The single celled organisms, the protozoa of which *Amoeba* and *Paramecium* are examples, represent the **unicellular level of organisation**. All the life processes are carried out within a single unit of protoplasm, although there may be considerable differentiation of the protoplasm into organelles; for example, in many fresh-water protozoa, organelles called contractile vacuoles are present; they are important in osmo-regulation and remove excess water which has entered by osmosis. The maximum size that can be reached by a unicellular organism is about $1\mu m$ in diameter. As the cell increases in size its volume gets very much bigger in relation to its surface area and a point is reached when the cell membrane is not large enough to take in enough food and oxygen for the increasing volume; at this point the cell may divide into two.

Primitive multicellular organisms have reached the **tissue level of organisation**. They have few organs and the life processes are mainly carried out by tissues and isolated cells. Such an organism is the coelenterate *Chlorohydra (Hydra) viridissima*; it lives in fresh-water ponds and is about 1 cm high when fully extended. It has a simple sac-like body with one opening which serves as a mouth and anus and which leads into the body cavity or enteron; below the mouth is a ring of tentacles. The body wall is arranged in layers, each one cell thick: there is an outer ectoderm and an inner endoderm, separated by a structureless jelly, the mesogloea. Most cells making up these layers are differentiated into musculo-epithelial cells which are integrated to form a tissue; they function in body movements. The muscle tails in the ectoderm run longitudinally and when they contract the body becomes short and squat, while the muscle tails in the endoderm run in a circular direction and when they contract the body becomes long and thin. Musculo-epithelial tissue is also used in locomotion by somersaulting and looping.

Musculo-epithelial cells next to the enteron are further differentiated into flagellar and pseudopodial cells. The flagellar cells help to keep the contents of the enteron moving, so aiding extracellular digestion and circulation; the pseudopodial cells engulf small particles of food and take them into the cells as food vacuoles for intra-cellular digestion. The nerve cells can be said to constitute a tissue, but gland cells and sensory cells occur in isolation between the musculo-epithelial cells and do not constitute tissues. Most of these cells do not work independently, however; for example, the sensory cells are connected to the nerve cells. Whether or not tentacles can be considered organs is debatable; they are composed largely of musculo-epithelial and nervous tissue and are co-ordinated for movement and for transferring the food to the mouth once it has been captured by the nematoblasts (page 60). Ovaries and testes are sometimes regarded as organs.

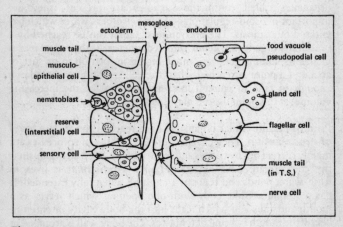

Figure 10. Structure of the body wall of Hydra *(L.S.)*

As a result of becoming multicellular, organisms can increase in size and complexity, and cells become specialised for specific functions, that is, division of labour occurs. This leads to greater efficiency and the opportunity to exploit a wider range of environments. However, it also means that cells are more dependent on each other and as organisms become larger other problems arise: for example, the movement of substances through the organism by diffusion is inadequate, and this has led to the development of transport systems to overcome this problem. Also, some means of co-ordinating the activities of the organism is required so that it functions as a whole.

Levels of organisation in plants

As in animals some plants consist of a single cell, and this functions as an organism, for example *Chlorella*. *Spirogyra*, which is a filamentous alga, consists of strands of identical cells, each cell able to function independently, with the exception of not being able to carry out sexual reproduction. Mosses, although differentiated into stem and leaves, still do not show much tissue specialisation. The cells differ mainly in shape, and little else. The flowering plants (Angiosperms) represent the most advanced plants, and have the highest degree of specialisation. They are differentiated into stem, root and leaves, each externally and internally specialised to carry out particular functions.

Structure of the stem The stem may be used to demonstrate how the plant tissues already described are grouped to form an organ. (The structure of the leaf is discussed in chapter 2; and the structure of the root in chapter 4.) The function of the stem is that of an organ of elongation. It separates out the leaves, exposing them to the light, and holds up flowers and fruits for pollination and dispersal respectively. It also transports water, mineral salts and soluble organic substances. In order to be able to carry out its function the stem must be firm enough to remain upright against mechanical forces, such as the wind, but also be flexible. The type and distribution of the tissues of the stem are important in giving the stem its essential properties. The structure of the stem of a perennial plant alters with age, since the weight it must support alters each year as a result of growth.

The stem in the case of an herbaceous plant is kept upright by a combination of factors. The parenchyma cells of the pith and cortex are normally fully turgid, and push out against the restraining

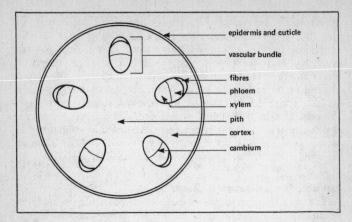

Figure 11. T.S. of young herbaceous stem

tension of the epidermis. The vascular bundles contain strengthening tissues in the form of xylem vessels and fibres. The vascular bundles, by being arranged in a cylinder around the stem, give flexibility as well as strength. In a very young stem the cambium is restricted to within the vascular bundles, but in any plant which produces a woody stem the cambium will extend and join up across the cortex. This cylinder of cambium will undergo cell division to produce more xylem inwards, and more phloem outwards. This results in an increase in the girth of the stem, and continues every year in the case of perennials. The stem will gradually change from a slender green herbaceous form to a woody trunk, characteristic of trees and shrubs.

The xylem formed by the cambium in spring, when growth is rapid, is different in appearance from the xylem formed in late summer as growth slows down. The demarcation between spring and summer xylem is called an annual ring. The age of a felled tree may be estimated by counting these rings. The extra xylem or wood, formed each year, is called **secondary thickening**. This activity of the cambium results in the epidermis being ruptured and pushed off. A layer of dividing cells, called the cork cambium, forms beneath the surface of the stem. The cork cambium divides to produce the bark which replaces the epidermis in forming a waterproof layer around the outside of the woody stem. The stems described are typical of land plants growing in conditions of adequate water supply

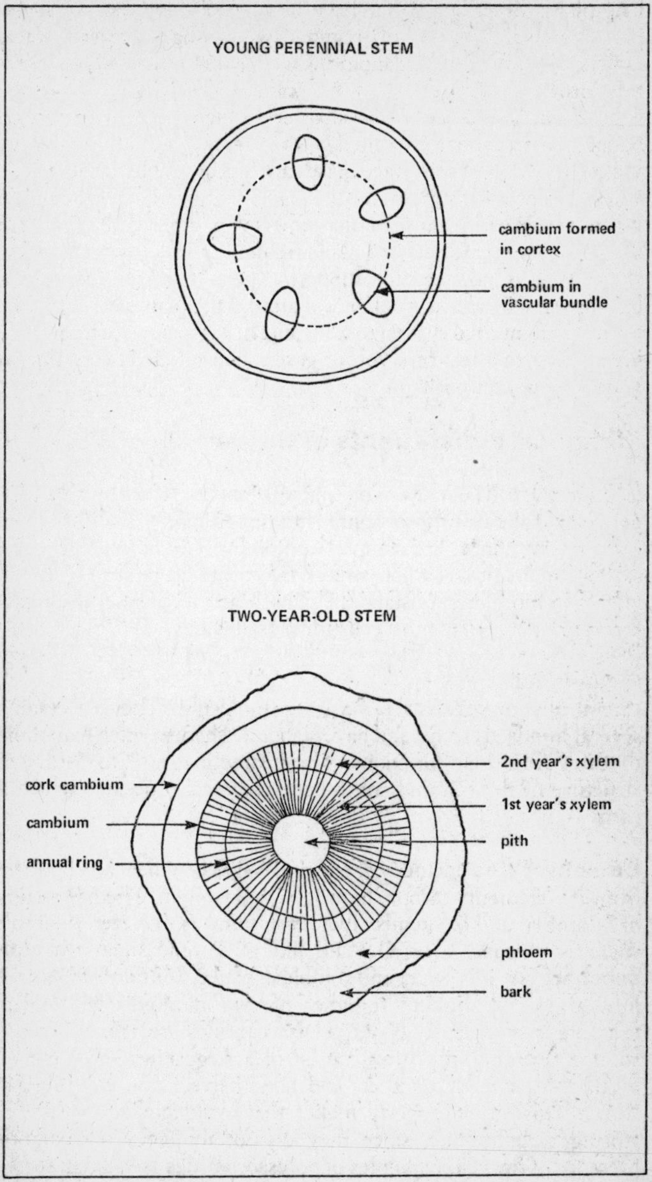

YOUNG PERENNIAL STEM

cambium formed
in cortex

cambium in
vascular bundle

TWO-YEAR-OLD STEM

cork cambium

cambium

annual ring

2nd year's xylem

1st year's xylem

pith

phloem

bark

Figure 12. Development of a woody stem

27

The plants described so far have been **mesophytes**. However, plants which survive in regions which have severe water deficiencies must have adaptations which allow them to cope with this problem. These plants are called **xerophytes**, and the adaptations are described in chapter 4. Plants which live in an aquatic environment are called **hydrophytes**, and although they will suffer no water shortage, gases are less available to them. The water will provide support for the plants living in it or on it, and so it is not surprising that their anatomy is very different from that of a mesophyte. A submerged plant frequently lacks sclerenchyma, since there is no need for support. Xylem tissue is reduced or lacking, since water is available all around the plant, obviating the need for an internal system to transport it. Air spaces are frequently present to provide a reservoir of gases. The cuticle is very thin as water loss is not a problem.

Chemical constituents of the cell

Chemically, cells consist of organic substances, which make up 20 per cent of the cell's total composition, and inorganic substances. Organic substances are complex compounds, all of which contain carbon, and include carbohydrates, fats, proteins, nucleic acids and vitamins. Inorganic substances include water and minerals; in fact at least 70 per cent of most cells consists of water.

Organic substances
Chemically, these consist of very large molecules. They can contain several hundred atoms, and have elaborate shapes which help them to carry out their functions. These functions are to form the structure of the cell and control the chemical processes which it carries out.

Carbohydrates include starch and sugars, which are built up from the elements carbon, hydrogen and oxygen. The proportion of hydrogen and oxygen is always the same as in water. A simple sugar is glucose, $C_6H_{12}O_6$, and like all simple sugars, **monosaccharides**, it is sweet and soluble in water. **Disaccharides** are formed by the linking together of two monosaccharides; an example is sucrose, $C_{12}H_{22}O_{11}$. Disaccharides are also sweet and soluble. **Polysaccharides** consist of a large number of monosaccharides joined together, and are unsweet and insoluble in water. This latter property makes them valuable as a means of storing carbohydrate, since they do not influence the osmotic pressure of the cell. Examples of polysaccharides are starch, stored in plant tissues, and glycogen, stored in animal tissues. The function of most carbohydrates is to provide energy for the cell. Examples of foods rich in starch are potatoes and cereals, and those rich in sugar are fruits and honey. Cellulose is an example of a carbohydrate which is an important structural compound. It forms the basis of all plant-cell walls, and like starch is made up of hundreds of glucose units joined together in a complex form.

Proteins are molecules which always contain the elements carbon, hydrogen, oxygen and nitrogen, and in addition sometimes phosphorus and sulphur. Proteins are built up from units called amino-acids joined together. A single protein may contain thousands of amino-acids. There are twenty different amino-acids which occur in living things. Some proteins contain a large proportion of the possible twenty, each present many times; whilst other proteins only contain a small proportion of the possible twenty amino-acids, each present many times. Proteins are needed for growth and replacement of tissues, but cannot be stored in the body. Examples of food rich in protein are eggs, meat and milk. An example of a structural protein is keratin, which forms the basis of hair and nails. Proteins also play a vital role in the metabolism of the cell, since all **enzymes** are proteins.

Enzymes are organic catalysts, which means that they speed up chemical reactions in organisms. Within a living cell up to a thousand chemical reactions may take place, but without the presence of enzymes these would proceed so slowly as to be hardly detectable. Most enzymes function within the cell where they are synthesised, and are called **intra-cellular**. Some enzymes pass out of the cell in which they were synthesised and are called **extra-cellular**. Examples of extra-cellular enzymes are digestive enzymes in mammals, and enzymes which fungi, such as *Mucor*, secrete on to the substrate on which they are growing. Enzymes are vital, not only because they speed up chemical reactions, but also because they control the direction of the chemical pathways within the cell. The presence or absence of a particular enzyme is under genetic control, and the method by which enzymes, like all proteins, are synthesised within the cell is described later. The fact that all enzymes are proteins is reflected in their properties.

Enzymes are damaged or denatured by high temperatures, and the graph in figure 13 shows the effect of temperature on a chemical reaction in a living cell. Up to 40°C an increase in temperature produces an increase in the rate of reaction, as in all chemical reactions.

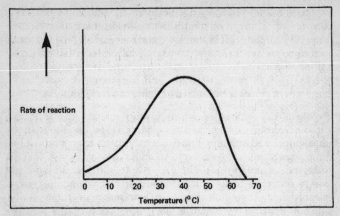

Figure 13. Enzyme activity and temperature

Above 40°C the rate of reaction begins to slow down with an increase in temperature. This is because the enzyme has been denatured and so is no longer able to catalyse the reaction. The reaction ceases completely around 60°C. The reaction takes place at its maximum rate at 40°C, known as the **optimum temperature** for that reaction.

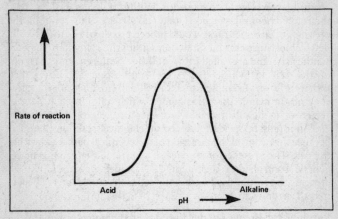

Figure 14. Enzyme activity and pH

Enzymes are very sensitive to the degree of alkalinity and acidity (pH) of their surroundings, and each enzyme can only function within a narrow pH range. The optimum pH for any enzyme varies considerably: for example, salivary amylase, which functions in the mouth, has an optimum pH of 7·5 (slightly alkaline); pepsin, which functions in the stomach, has an optimum pH of 2·0 (acid).

Each enzyme will act only on one substance or type of substance; for example, salivary amylase will act only on starch, urease only on urea. This is called enzyme **specificity**. The end products of a chemical reaction involving an enzyme are the same as when the enzyme is not present. This is because an enzyme can only speed up a chemical reaction. At the end of the reaction the enzymes are unaltered and so they may be re-used. (In practice enzymes will eventually have to be replaced as they may, for example, be carried away in the gut.) Enzymes are vital to life, and conditions which prevent enzymes functioning will also kill or inactivate the organism. The lethal effect of some chemicals, such as cyanide, is caused by their preventing the action of an enzyme.

Fats consist only of the elements carbon, hydrogen and oxygen, but the oxygen is present in smaller proportions than in carbohydrates. The term **lipids** is given to fats and oils, the only difference between the two being that oils are liquids at room temperature, and fats are solids. Fats are made up of two sorts of units joined together, fatty acids and glycerol. Fats are used as a store of energy in organisms. This is because fats are insoluble and compact, and per unit of weight yield more energy than any other food. In all mammals fat acts as an insulating layer beneath the skin, but in aquatic mammals the amount of fat is much greater, because aquatic mammals cannot use air trapped amongst hair as an insulating layer. Blubber, for example, is in fact a whale's vast supply of insulating fat. Foods rich in fat are butter, cooking oils, milk and chocolate. Excess carbohydrates in our diet are converted into fats and stored.

Vitamins are complex organic substances which are needed in the diet in small quantities. They carry out a large range of functions, and the severe shortage of any vitamin may result in a characteristic deficiency disease (see page 227). Plants are able to synthesise their own vitamins, but animals require most of them to be present in their food.

Nucleic acids are long chain molecules which occur in all cells and organisms, including viruses and bacteria. Their universal presence in living cells, in some form, gives an indication of the vital roles they perform. There are two types of nucleic acids, deoxyribonucleic acid (DNA) which occurs only in the nucleus, and ribonucleic acid (RNA) which occurs mainly in the cytoplasm, but also in small quantities in the nucleus. The structure of DNA was established by two famous research scientists, Watson and Crick, working in Cambridge in 1953. DNA is the chemical substance which makes up the genetic information. DNA is a long thread-like molecule which always exists as two chains joined to each other. The DNA molecules are not straight but curve round like a spiral staircase, and have been given the name 'the double helix'. DNA is carried in the nucleus of every cell, and along with proteins it forms structures called **chromosomes**. Each DNA molecule consists of a backbone made up from sugar (deoxyribose), and phosphate molecules. There is no variation in this backbone, either from cell to cell or from organism to organism. The links joining the sugar-phosphate backbones together are made up from four possible molecules called **bases**. These bases are called adenine (A), thymine (T), guanine (G) and cytosine (C). Each base is attached to the sugar molecule of the DNA backbone forming a side chain which faces inwards. It is these bases which hold the two backbones together to form the 'double helix'. The bases pair only with one another in a specific way: adenine pairs with thymine, and guanine pairs with cytosine. This is because these combinations occupy the same space, and so the backbones of the helix are the same distance apart.

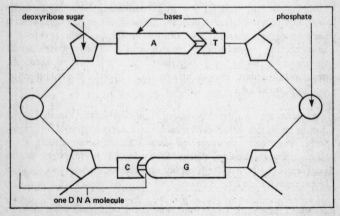

Figure 15. Structure of DNA

The bases are held together only by weak chemical bonds. The structural arrangement of the DNA molecule allows it to carry out two essential functions efficiently: firstly it can replicate (make an identical copy of itself), and secondly it can carry an almost infinite amount of information for the cell, in a coded form. Replication will be discussed in this chapter, but the code will be discussed in chapter 10, on genetics. Each time the cell divides, the genetic material makes a copy of itself, so that the two new daughter cells both carry identical sets of information. In order to do this the DNA unwinds, and the two chains separate by means of the breaking down of the bonds holding paired bases. Then a new correct base is paired up with each base on the parent chain, and the bonds form again. In this way two DNA chains are formed.

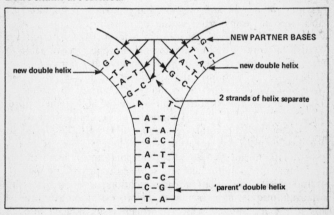

Figure 16. Replication of DNA

Ribonucleic acid (RNA) is chemically very similar to DNA, but differs in one or two respects. Firstly it exists only as a single chain, and secondly the base thymine is replaced by the base uracil in RNA. RNA plays a vital role in protein synthesis, and so moves from the nucleus into the cytoplasm, where it is closely associated with the ribosomes.

Inorganic substances
Minerals, like vitamins, are only needed in small quantities by organisms, but unlike vitamins are chemically simple. They fulfil a variety of functions in the cell, and a summary of them may be found on page 226.

The role of **water** in living organisms cannot be over-emphasised. About 60 per cent of most animal bodies consists of water, and up to 90 per cent of many plants. It is thought that life originally evolved in water, and many organisms are still totally aquatic. Even in terrestial organisms it is the environment in which all chemical reactions must take place. Dry tissues, such as those in seeds or dormant spores, are more resistant but metabolically inactive and will only resume the processes of life when their water content rises back to normal. The importance of water to all living things is a result of its properties.

Water is often referred to as the universal solvent, meaning that many things will dissolve in it easily, forming a solution. This property is of great importance for a number of reasons. All enzymes require an aqueous solution in which to function; all substances are transported within organisms in solution. Also water has a huge heat capacity, which means that it requires a lot of heat to raise its temperature. This property is valuable to living things since it helps to maintain them at a fairly constant temperature. In mammals the additional method of relying on the evaporation of water from the body's surface to produce a cooling effect is also used to help regulate the body temperature. Water also has the unusual property of becoming less dense below about 4°C. For this reason, ice floats on water, and so freezing begins at the surface of water where the temperature is lowest. This means that aquatic organisms can remain active at the bottom of a pond or lake and survive during cold periods. In addition to the functions already mentioned, water serves an important purpose as a reactant in the process of photosynthesis; this is the method by which hydrogen is introduced into all organic compounds. Water is also important in supporting plants, by maintaining the turgor pressure of their cells. Finally, water plays a vital role in reproduction because, with a few exceptions, male and female gametes unite and the zygotes develop either in water or in a fluid composed largely of water.

The movement of substances

From the description of the cell and its contents so far, it is apparent that substances are constantly moving within the cell, and also into and out of the cell. This is true whether the cell constitutes a unicellular organism or is part of a multicellular organism. There are several mechanisms involved in movement at the cellular level, but these are usually efficient only across short distances. As organisms increased in size, transport systems evolved to deal with the problem of movement across greater distances than just one cell. Two processes which, though physical processes, are important means of transport in cells, are diffusion and osmosis.

Diffusion is defined as the movement of molecules from a high concentration to a low concentration. This movement will continue until the two concentrations are equal. Diffusion will occur only in gases or liquids where the molecules are spaced out sufficiently to carry out their natural movements. When diffusion occurs in cells any membranes separating the unequal concentrations of molecules must be fully **permeable** to that particular molecule. This means that the molecule must be able to pass freely through the membrane as if it were not present. Membranes permeable to a particular molecule may be thought of as having spaces which are larger than the molecule. Plasma membranes and cellulose cell walls are fully permeable to water, and to gases such as carbon dioxide and oxygen.

Osmosis, another physical process, is in fact a special type of diffusion, and is defined in biology as the movement of **water** from a weak solution to a strong solution through a **semi-permeable membrane**. Osmosis occurs when a membrane or partition is permeable only to the water molecules of a solution, but other molecules dissolved in the water are too large to pass through it. This type of membrane is called semi-permeable, and applies to the plasma membrane but not to the cellulose cell wall. This means that in living systems water is free to move in or out of cells, but large molecules, such as sugars, are unable to pass through the plasma membrane. Water will always be able to move from high concentrations to low concentrations, i.e. from a weak solution (with a high concentration of water molecules) to a strong solution (with a low concentration of water molecules).

The process of osmosis may be demonstrated in a non-living system using the apparatus shown in figure 17. The result of this experiment is that the sugar solution rises up the capillary tube and becomes more dilute. This may be explained by the movement of water molecules from a high concentration of them, in the pure water, to a low concentration of them, in the sugar solution, through the cellophane. The solution rises up the tubing owing to a pressure. This pressure is a result of the strong sugar solution taking in water molecules and is known as the **osmotic pressure** of the solution. The size of the osmotic pressure is determined by the difference in the concentrations of the solutions on either side of the semi-permeable membrane. In the experiment described, if the sugar solution had been outside the thistle funnel and the water inside it, water would have moved out of the funnel into the beaker.

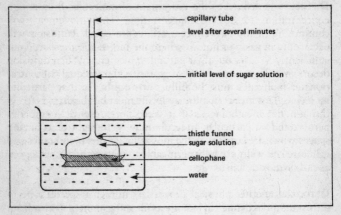

Figure 17. Demonstration of osmosis

Animal cells have a semi-permeable membrane (the plasma membrane) separating their contents from the environment. As a result any changes in the osmotic concentration of the cell's environment will result in water either entering or leaving the cell. In order to prevent this damaging the cell the external conditions must be controlled, as the blood plasma is for the cells of the body, or the cell must have some method of controlling its water content. This is the function of the contractile vacuole in *Amoeba*, and these methods of **osmo-regulation** are discussed more fully later in the book.

Plant cells are separated from their environment by the cell wall, a dead boundary which is fully permeable but fairly inelastic. Within the cell wall lies the plasma membrane which represents the living boundary of the cell. The vacuole of the cell contains a solution of sugars, salts, etc., which is contained by the semi-permeable plasma membranes. When a plant cell is surrounded by a weak solution water will tend to move into the vacuole by osmosis. This results in a limited increase in the size of the cell contents. The cell does not rupture because of the presence of the cell wall, which is mechanically strong and can resist osmotic pressure. When the pressure of the vacuole, pushing outwards, is equalled by the resistance of the cell wall, the cell will be unable to absorb any more water by osmosis, and is described as fully **turgid**. Herbaceous plants rely on this turgidity to support their stems, and most plants rely on it to hold out their leaves. In normal conditions the movement of water into a plant cell is a balance between two forces; firstly the osmotic pressure (O.P.) of the cell sap, tending to draw water into the vacuole, and secondly the resistance of the cell wall (wall pressure, W.P.), tending to force water out. The balance between these two forces it called the suction pressure (S.P.) and represents a cell's ability to take in water. The relationship between these forces may be represented by the equation S.P. = O.P. − W.P. When a cell is fully turgid the osmotic pressure is equal to the wall pressure, and the suction pressure is zero, and so the cell has no tendency to absorb water. In laboratory conditions it is possible to immerse a plant cell in a solution with a higher osmotic pressure than that of the cell vacuole. Under these conditions water will move out of the cell by osmosis from the higher concentration in the vacuole to to the lower concentration of water in the external solution. This results in the vacuole shrinking in size and the cytoplasm falling away from the cell wall. This phenomenon is known as **plasmolysis**. When a plasmolysed cell is placed in water it exerts its maximum suction pressure since the wall pressure is zero. The suction pressure is in fact equal to the osmotic pressure of the cell.

As well as the physical processes of osmosis and diffusion, organisms employ other methods of moving substances. Some of these methods have only recently been investigated and there is still a great deal to find out about them. **Active transport** is an example of a method by which a cell uses energy to drag in substances against the diffusion gradient. Many plant cells, for example, maintain certain substances within their vacuoles at concentrations many times higher than that of their surroundings and it would be expected that these substances would diffuse out of a cell. It has been proved that the cell uses energy to keep substances at high concentrations because anything that interferes with respiration influences the rate at which active transport takes place. Cells also use active transport to pump substances out. Particles may enter a cell by the plasma membrane engulfing them and forming a vacuole. This process is called **phagocytosis**.

Cell division

Cell division occurs during the growth and reproduction of an organism. The most important structures involved in cell division are the chromosomes, present in the nucleus. Depending on the behaviour of the chromosomes, two types of cell division may be recognised, **mitosis** and **meiosis**. Chromosomes are fine strands consisting chemically of deoxyribonucleic acid (DNA) and protein. Each organism has a characteristic number of chromosomes in each cell; man, for example, has 46 chromosomes, the garden pea 14 and the shrimp 254. In the majority of plants and animals these chromosomes exist in pairs, so it would be more accurate to think of man, for example, as having two sets of 23 chromosomes in each nucleus (the members of each pair are called homologous chromosomes). This state of having two sets of chromosomes is called **diploid**. When a diploid cell undergoes mitosis, the two daughter cells produced are also diploid. This type of division occurs during growth and asexual reproduction. When a diploid cell undergoes meiosis the daughter cells produced contain one set of chromosomes (half the number of the original parent cell) and this state is known as **haploid**. Meiosis must occur at some stage of the life cycle of organisms carrying out sexual reproduction; it keeps the number of chromosomes in successive generations constant. Meiosis in man is illustrated in figure 18.

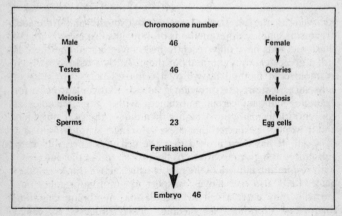

Figure 18. Meiosis in man

If each egg cell and sperm contained 46 chromosomes, the child produced would contain 92 chromosomes (twice the number of its parents). It is important that each cell contains an accurate copy of the chromosomes since they constitute the genetic material and carry in a coded form all the information the cell requires for its development and functioning.

Mitosis is a process broken down for convenience into a number of stages – prophase, metaphase, anaphase, and telophase – although it is in fact a continuous process. The process takes between one to three hours to complete, and the 'resting' time between successive cell divisions is called **interphase**. A short interphase time, indicating rapid cell division, would be between 10 and 24 hours. Interphase is necessary because the cell carries out all the preparations for cell division in this time. The chromosomes duplicate themselves to provide sufficient DNA for the next two daughter cells, and structures such as chloroplasts, mitochondria and centrioles are newly formed. The centrioles always exist in pairs and are rod-shaped structures which only occur in animal cells. In these cells the movements of centrioles are an essential part of cell division. All the phases of mitosis are shown together in figures 19 and 20.

During **prophase** the chromosomes become visible owing to their coiling up. This makes them compact structures which can move about the cell more easily. The nuclear membrane and nucleolus break down and the centrioles separate. As the centrioles separate and take up their position at opposite ends of the cell, a fibrous structure forms between them called the spindle. As prophase continues the chromosomes appear as double structures held together by the region called the centromere. The two parts of the chromosome are called chromatids and are genetically identical, produced by the duplication of the DNA. The centrioles take up their final position establishing the poles of the spindle.

In **metaphase** the chromosomes attach themselves by their centromeres to the spindle and position themselves in the central region of the spindle, called the equator. Next, in **anaphase**, the two halves of chromosomes spring apart and appear to be pulled to the nearest pole. This happens simultaneously in all the chromosomes, the centromere being in the lead, taking its chromatid with it. The mechanism of this movement is still not satisfactorily explained. In **telophase** chromatids cluster tightly around the poles and now represent the chromosomes of the daughter cells.

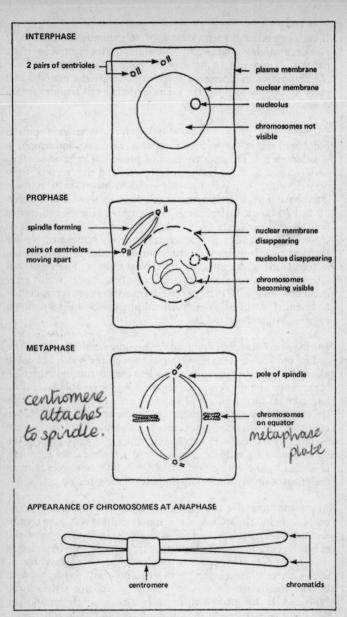

Figure 19. Stages in mitosis (1)

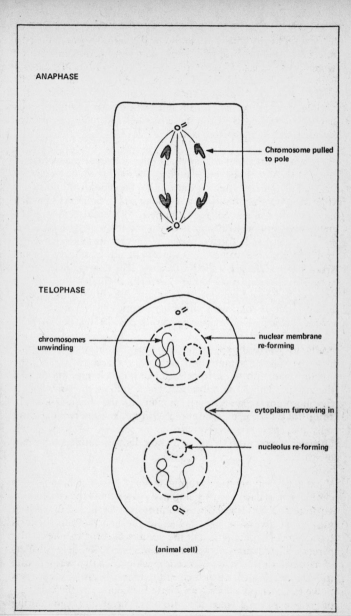

Figure 20. Stages in mitosis (II)

The nucleolus and nuclear membrane begin to re-form in the two new cells and the chromosomes begin to uncoil. The method by which the cytoplasm is separated into two cells differs in plant and animal cells. In animal cells the cytoplasm constricts until two cells are pinched off. In plant cells a new cell wall forms across the middle of the cytoplasm. As a result of mitosis two daughter cells have been formed with the same number and types of chromosomes as the original parent cell.

Meiosis takes place in two main stages, called meiosis I and meiosis II. As a result four daughter cells are produced from the original parent cell. Meiosis I and II are broken down into stages (like mitosis). **Prophase I** is a very long and complex stage, which is represented in a simplified form here. As in mitosis the nuclear membrane and nucleolus break down and the centrioles separate, forming the spindle between them. The chromosomes condense and homologous chromosomes come to lie side by side forming an arrangement which is called a bivalent. The chromatids become wrapped around each other as further condensing of the chromosomes occurs. At certain points the chromatids become broken and re-join with other chromatids. This results in a characteristic cross-formation appearing, called a chiasma.

Next, in **metaphase I**, homologous chromosomes become attached to the spindle and position themselves around the equator, one centromere from each pair lying on opposite sides of the equator. During **anaphase I**, which follows, the homologous chromosomes separate, each moving to the nearest pole. In **telophase I** the cytoplasm often separates and the two daughter cells may pass into interphase, as in mitosis. In some cases, however, there is no delay, and the cells pass immediately into meiosis II.

What has been achieved during meiosis I is that homologous chromosomes have been separated. In **meiosis II** the chromatids separate in a very similar way to mitosis. The second division of meiosis is always at right angles to the first division. Meiosis results in cells containing half the number of chromosomes as the parent cell, and also cells which are not genetically identical. This variation in the four daughter cells is achieved in two ways: firstly, through it being chance which of the homologous chromosomes move to which pole during anaphase I; secondly, by the exchange of genetic material when the chromatids break and rejoin during prophase I.

42

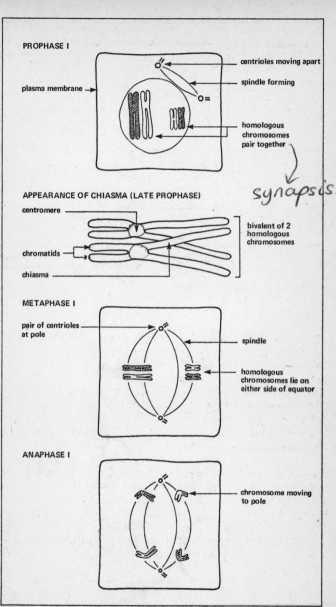

PROPHASE I

centrioles moving apart

spindle forming

plasma membrane

homologous chromosomes pair together

synapsis.

APPEARANCE OF CHIASMA (LATE PROPHASE)

centromere

bivalent of 2 homologous chromosomes

chromatids

chiasma

METAPHASE I

pair of centrioles at pole

spindle

homologous chromosomes lie on either side of equator

ANAPHASE I

chromosome moving to pole

Figure 21. Stages in meiosis (I)

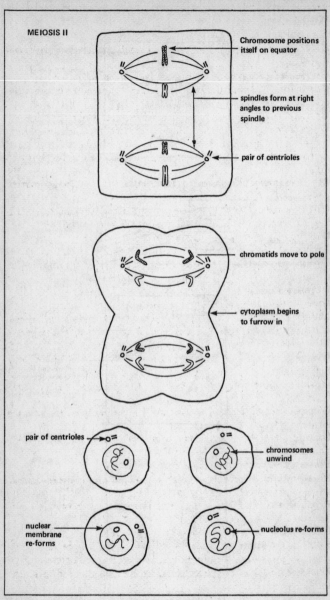

Figure 22. Stages in meiosis (II)

44

Key terms

Cell Unit of protoplasm containing a single nucleus bounded by a membrane.

Cell wall Dead cellulose layer surrounding plant cells.

Centriole Paired cylindrical structures which determine the position of the spindle in animal cells.

Centromere Part of chromosome which attaches it to spindle.

Chloroplast Contains chlorophyll and is the site of photosynthesis in plant cells.

Chromatid One of the two strands produced when a chromosome duplicates.

Chromosome Thread-like structure present in nucleus; carries genes.

Diffusion Movement of molecules from a high concentration to a low concentration (until the two concentrations are equal).

Diploid Having two sets of paired chromosomes.

Enzyme A catalyst produced only by living cells.

Homologous chromosomes Corresponding chromosomes of the same shape, one obtained from each parental gamete.

Haploid Having one set of unpaired chromosomes.

Inorganic substances Of mineral origin.

Meiosis Cell division in which the chromosome number is halved.

Mitochondrion Organelle present in all cells, site of respiration.

Mitosis Cell division producing two identical cells.   6

Nucleus Structure in a cell which controls its development and functioning.

Organ Collection of tissues which are specialised in structure and function.

Organic substances Complex carbon compounds, many only formed in metabolism.

Osmosis Movement of water from a weak solution to a strong solution through a semi-permeable membrane.

Plasma membrane Living boundary in all cells.

Plasmolysis Movement of water out of cell when immersed in a solution stronger than the vacuole.

Protoplasm The living material of cells, differentiated into cytoplasm and nucleus.

Ribosomes Site of protein synthesis in all cells.

Semi-permeable membrane Allows small molecules, e.g. water, to pass through more easily than large molecules, e.g. sugar.

Tissue Collection of similar cells which are specialised in structure and function.

Chapter 2
Nutrition

Nutrition is a characteristic of all organisms. It is the method by which the organism obtains organic substances. These substances are used for growth, as a source of energy, or for producing complex molecules, such as enzymes. An organism's method of nutrition determines many other of its features. Different methods of nutrition are, for example, responsible for all the other basic differences between plants and animals, and for many of the adaptations of a parasite, such as the tapeworm. The nutritional methods of organisms such as bacteria and fungi have meant that man has had to resort to techniques such as freezing and pasteurisation to prevent spoilage of food. There are two basic types of nutrition:

Autotrophic nutrition

This involves the synthesis (building up) of organic compounds from inorganic raw materials. It is often called holophytic nutrition and it is carried out by most plants. Most autotrophic organisms build up organic compounds from carbon dioxide and water, and use light energy, trapped by chlorophyll, to drive the process. This process is called **photosynthesis**.

Heterotrophic nutrition

This is the method of feeding characteristic of all animals and fungi, and most bacteria. These organisms are unable to synthesise their own food and so obtain it in a complex organic form. These organisms are therefore directly or indirectly dependent on green plants which have produced these organic substances from inorganic materials. There are three types of heterotrophic nutrition.

Holozoic nutrition involves feeding on the solid organic material from plants and animal bodies. This method is used by animals.

Saprophytic nutrition involves feeding off the soluble organic compounds from the remains of plant and animal bodies. It is used by many bacteria and fungi, some of which live in the soil and are responsible for the decay of plant and animal bodies. This results in the release of elements which can be reused by green plants.

Parasitic nutrition involves feeding off the organic compounds present in the body of another living organism. It is used by some animals, a few flowering plants, some bacteria and fungi, and all viruses. These parasitic organisms may cause diseases.

These types of nutrition are now considered in more detail.

Autotrophic nutrition

The process of photosynthesis is of vital importance to all living things since it is the only method by which organic materials are produced. It has been estimated that an acre of corn can produce 10,000 kg of sugar per year. During photosynthesis light energy is converted into chemical energy in the form of sugars. Sugars are stable compounds which all organisms can use as a source of energy. Once the energy has been released, during respiration, for warmth or movement, that energy is lost from the organism and more food must be obtained to provide more energy. There is therefore a constant flow of energy through the plant and animal kingdoms originating from the sunlight. The significance of photosynthesis can be seen clearly in the food webs on page 207, and also in the carbon cycle on page 208.

Photosynthesis can be represented by the equation

$$6CO_2 + 6H_2O \xrightarrow[\text{chlorophyll}]{\text{light energy}} C_6H_{12}O_6 + 6O_2$$

(carbon dioxide + water → glucose + oxygen)

This equation indicates the raw materials needed for photosynthesis and the initial products of the process. Experiments can be carried out to demonstrate the conditions necessary for photosynthesis to take place and to demonstrate the products of the process. Two points which the equation fails to make, and which will be discussed later, are that it is not only sugars which are produced as a result of photosynthesis but the carbon skeletons for carbohydrates, fats and proteins too; and that photosynthesis takes place in a large number of steps, each step requiring a specific enzyme. The details of the mechanism of photosynthesis were only established in the later half of this century. This was a result of more sophisticated techniques being developed in other scientific fields, such as the use of radioactive isotopes.

The site of photosynthesis Photosynthesis may occur in any part of the plant containing chlorophyll, such as a green stem, but it is carried out mainly by the leaves in the majority of plants. The leaves of a typical **mesophytic** plant are well adapted to allow photosynthesis to occur efficiently. The entire process of photosynthesis occurs inside the chloroplasts, and so it is essential for the raw materials of the process to reach the chloroplasts, and for the products to move away easily. The features of the leaf which allow these exchanges to occur may be summarised as follows.

1. The leaf is thin and flat, and this produces a large surface area which is exposed to the sunlight and carbon dioxide.
2. The thinness of the leaf means that gases have only to diffuse across short distances.
3. There are a large number of gaps, or stomata, on the lower surface of the leaf which allow gaseous exchange between the air in spaces inside the leaf and the outside atmosphere.
4. The cells of the spongy mesophyll are loosely packed and the presence of air spaces allows gases to diffuse easily between cells.
5. The chloroplasts are more numerous in the palisade cells, which are beneath the upper epidermis, where there will be more sunlight.
6. The elongated shape and the position of the palisade cells minimises the number of cross walls which will absorb the sunlight before it reaches the chloroplasts.
7. A network of veins serves as a mechanical framework for the delicate leaf. The veins also ensure that every cell is close to a xylem vessel and phloem tube. The xylem vessels transport an adequate supply of water and mineral salts to the cells of the leaf, and the phloem tubes transport away the sugars produced during photosynthesis.
8. Leaves are spread out along the stem in regular patterns to minimise over-shadowing.

Unfortunately many of the features listed also allow rapid water loss by evaporation from the moist cell walls of the cells of the leaf. This evaporation is called transpiration and although it cannot be prevented when the leaf is photosynthesising various devices have evolved to minimise water loss. Stomata are of particular significance both for regulating water loss and also for being involved in controlling the rate of photosynthesis. When the stomata close during the day, for example, as a result of water shortage, the rate of photosynthesis will be much slower. Photosynthesis in these conditions will depend on the carbon dioxide released from respiration, and the two processes will be proceeding at about the same rate. When the stomata are open in bright light photosynthesis may proceed twenty times faster than respiration.

Photosynthesis actually occurs within the chloroplasts. Carbon dioxide will diffuse through the moist walls of the mesophyll cells, from the inter-cellular spaces of the leaf, and so to the chloroplasts. Water reaches the cells of the mesophyll by osmosis from the network of veins spreading throughout the leaf. These veins contain xylem vessels which are part of a continuous conducting system which transports water from the soil up through the roots and stem to the leaves. Energy to drive the process of photosynthesis is obtained from the light energy absorbed by the chlorophyll contained within the chloroplasts. Oxygen is produced during photosynthesis and will diffuse out through the moist cell walls into the inter-cellular spaces of the leaf. Starch may accumulate in the chloroplasts but enzymes eventually convert it into sucrose, and this is translocated through the phloem to all parts of the plant.

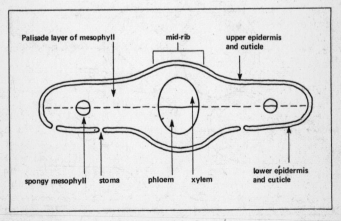

Figure 23. Transverse section through leaf

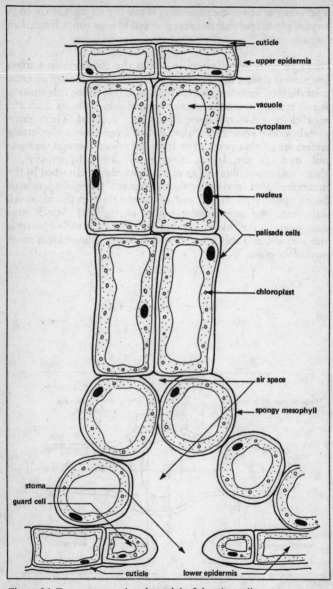

Figure 24. Transverse section through leaf showing cells

Experiments to show the conditions required for photosynthesis

Glucose is the main product of photosynthesis, but this may be built up into proteins for growth, broken down into carbon dioxide and water for energy production, or converted into starch for storage. When photosynthesis is occurring rapidly starch will accumulate in the leaf. Starch is insoluble and so, unlike sugars, will not affect the osmotic pressure of the cells. The formation of starch is therefore a useful way of establishing whether a plant is carrying out photosynthesis. In the dark all the starch is converted back into sugar and translocated through the phloem to other parts of the plant. In order for photosynthesis to occur a plant must have an adequate supply of carbon dioxide (CO_2), water and light, chlorophyll must be present and the temperature must be suitable. The necessity of these factors may be demonstrated by experiments on plants which have been de-starched by placing them in the dark for a minimum of twenty-four hours. In order to test for starch a leaf is first boiled in water and then alcohol to remove any pigments and so make it colourless. When placed in dilute iodine any parts containing starch will turn dark blue.

To show light is needed for photosynthesis

Part of a de-starched leaf is covered with a strip of opaque paper and left in bright light for several hours.

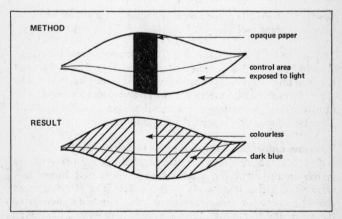

Figure 25. Method and result of experiment to show light is needed for photosynthesis

When the leaf is tested with iodine only the uncovered area turns dark blue indicating the presence of starch. This shows that light is essential for photosynthesis.

To show CO$_2$ is needed for photosynthesis

The leaves of the de-starched plant are placed in flasks as shown in figure 26.

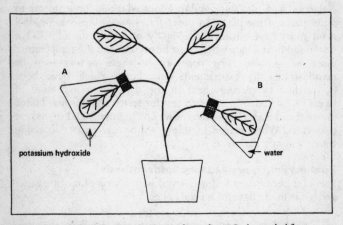

Figure 26. Method for experiment to show that CO$_2$ is needed for photosynthesis

The leaf in flask A is deprived of CO$_2$ since potassium hydroxide absorbs the gas. Flask B is the control experiment. The experiment is left in bright light and kept well watered for several hours. The leaves from the flasks are then removed and tested for starch. The leaf from flask A remains colourless, that from flask B turns dark blue. This shows CO$_2$ is needed for photosynthesis.

To show chlorophyll is needed for photosynthesis

A de-starched plant which has variegated leaves (e.g. privet or ivy) is left in the light for several hours. Variegated leaves lack chlorophyll either completely or in patches. The distribution of chlorophyll in a particular leaf is recorded and the leaf is then tested for starch. It is found that starch is only present where chlorophyll is present. This proves that chlorophyll is essential for photosynthesis.

To show that a suitable temperature is needed for photosynthesis

In order to measure the rate of photosynthesis the volume of oxygen released from a plant in a given time is recorded. A water plant is used because the bubbles of oxygen are easily collected from water. The apparatus used for this experiment is as in figure 27. This apparatus is placed in a water bath so that any given temperature is kept constant throughout the experiment. The result is that most oxygen is collected in a given time between 25°C and 35°C. Below 25°C the rate of oxygen release slows down and above 40°C no oxygen is collected since the enzymes involved are damaged. This shows that photosynthesis will only proceed efficiently within a certain temperature range.

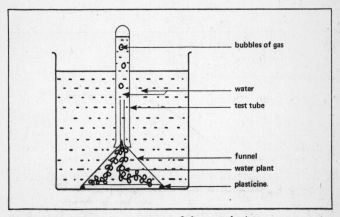

bubbles of gas

water

test tube

funnel
water plant
plasticine

Figure 27. Apparatus to measure rate of photosynthesis

To show water is needed for photosynthesis

It is impossible to demonstrate by simple experiments that water is needed for photosynthesis, since plants deprived of water will die because water is essential to several different aspects of plant life besides photosynthesis. Elements known as radioactive isotopes can now be made, and the use of these has shown that water is directly needed for photosynthesis. Radioactive isotopes are unstable elements which break down, releasing radioactivity, but which are treated by organisms as the normal elements. The pathway of these isotopes may be traced inside organisms by the release of radioactivity (detected by a Geiger counter), or by the ability of the isotopes to turn a photographic plate black. The experiments proving that water is needed for photosynthesis are described later along with other experiments involving the use of radioactive isotopes.

Interaction of factors affecting photosynthesis As the above experiments have shown, several factors are needed for photosynthesis to take place. The rate of photosynthesis may be slowed down if any one of these factors is in inadequate supply. The factor in shortest supply, and therefore controlling the rate of photosynthesis, is called the **limiting factor**. The principle of limiting factors may be illustrated by placing a plant in increasing light intensities and measuring its rate of photosynthesis, whilst keeping the temperature and CO_2 concentration constant.

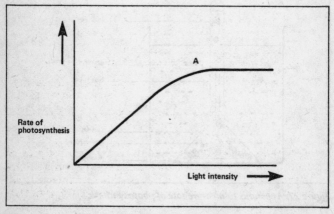

Figure 28. Graph showing effect of light intensity on photosynthesis

The results shown on the graph indicate that up to A an increase in light intensity produces an increase in the rate of photosynthesis, but after that point the rate levels off. This levelling off could be due to several reasons: the temperature may not allow the chemical changes to proceed any faster; there may be an inadequate supply of CO_2; the products of the reaction may be building up, preventing any further action; or this may be the maximum rate at which the 'photosynthetic machinery' can work under any conditions.

In order to find out whether temperature or CO_2 concentration was acting as a limiting factor, the experiment would be repeated first at a higher temperature, and secondly with a higher CO_2 concentration. Assuming that the CO_2 concentration was acting as the limiting factor, the result shown in figure 29 would be obtained. In this case increasing the amount of CO_2 allows photosynthesis to reach a higher maximum rate.

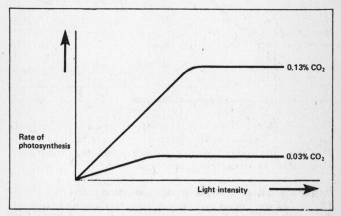

Figure 29. Photosynthesis, light intensity, and CO_2 concentration

It is essential for active growth in plants that the rate of photosynthesis (the production of carbohydrates) exceeds the rate of respiration (the breakdown of carbohydrates). When the two processes take place at the same rate the **compensation point** has been reached. The balance between these two processes in relation to light intensity can be represented by figure 30. The graph shows the balance between **two** processes, respiration and photosynthesis. A green plant (like all living things) will respire all the time (both in the light and in the dark), **absorbing** oxygen. In the light the plant will also carry out photosynthesis, the amount of oxygen **released** being related to the light intensity. At point A, in the dark, only respiration occurs, and oxygen is being absorbed. During stage B photosynthesis begins to occur, as the light intensity increases, and some oxygen is released. At point C, **the compensation point**, respiration and photosynthesis are taking place at the same rate, and so the same amount of oxygen is being absorbed as is being released.

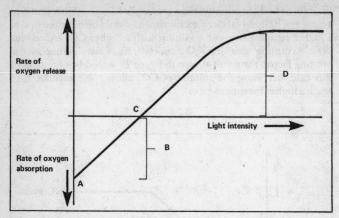

Figure 30. Graph showing compensation point

At the point C there is no net gain or loss of oxygen. During stage D the rate of photosynthesis continues to increase with increasing light intensity, and is taking place more rapidly than respiration. This leads to an overall release of oxygen.

Experiment demonstrating compensation point

The rate of respiration may be measured by showing how much CO_2 is released in a given time. The rate of photosynthesis may be measured by showing how much CO_2 is absorbed in a given time. The removal or addition of CO_2 to an atmosphere may be shown by using a substance called bicarbonate indicator solution. This changes colour depending on its acidity or alkalinity.

Colour Purple   Red   Orange   Yellow
pH   Alkaline ◄─────────────► Acid

Since CO_2 is an acid gas (when it dissolves in water it produces carbonic acid), the removal or addition of CO_2 from bicarbonate indicator causes it to change colour. The addition of CO_2 changes the colour towards the yellow end, and the removal of CO_2 changes the colour towards the purple end.

The experiment is set up as in figure 31. A, B, and C are corked test tubes containing equal-sized leaves and equal amounts of bicarbonate indicator. A is uncovered, so the leaf is exposed to light, B is covered with muslin, so the leaf obtains dim light, and C is covered with foil, so the leaf is in the dark. A, B, and C would each have a control, differing only in that it did not contain a leaf. D is a corked test-tube containing just the same amount of indicator as A, B, and C. D is not a control, but is being used as a comparison which will be referred to at the end of the experiment. All the tubes are kept in a water bath, so that they remain at a constant temperature. The water bath is brightly illuminated and the tubes are left there for several hours.

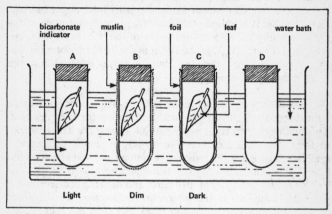

Figure 31. Exchange of CO_2 by green leaves

The colour of the indicator in A, B, and C is then compared with that of D. It is found that the indicator in A is more purple, in B it is the same colour, and in C it is more yellow. All the controls remain the same colour as D. This shows that in A CO_2 has been removed from the atmosphere, in B the CO_2 has stayed the same, and in C CO_2 has been added to the atmosphere. We can infer from this that in A photosynthesis has proceeded faster than respiration, in B they have proceeded at the same rate, and in C only respiration has occurred.

Chemical changes during photosynthesis
As mentioned earlier in the chapter, the overall equation for photosynthesis gives no indication of the complexity of the reactions involved. Each reaction requires a specific enzyme, and all these enzymes are localised within the chloroplast. Photosynthesis occurs in two main stages, the light and dark reaction.

Light reaction During this stage chlorophyll absorbs light energy and changes it into chemical energy. This energy results in water being split into hydrogen and oxygen. The oxygen is released as a waste product. The hydrogen is used up in the following dark reaction. Direct evidence that the oxygen released from photosynthesis comes from water and not CO_2 can be obtained by carrying out experiments using radioactive isotopes. This experiment also demonstrates that water is required for photosynthesis. The radioactive oxygen isotope is represented by $^{18}O_2$. Two experiments are carried out using the unicellular alga *Chlorella*. First the alga is supplied with the $^{18}O_2$ in water but none in the CO_2. Secondly the alga is supplied with the $^{18}O_2$ in the CO_2 but none in the water (H_2O). The result in the first experiment is that $^{18}O_2$ is released from the plant, but no $^{18}O_2$ is released from the plant in the second experiment.

Dark reaction As the name implies, this stage does not require light directly, only hydrogen produced during the light reaction. This hydrogen then combines with CO_2 during the dark reaction, to form carbohydrates. The role of CO_2 in photosynthesis was established by using the radioactive isotope of carbon represented by ^{14}C. Plants are allowed to carry out photosynthesis surrounded by $^{14}CO_2$ for varying lengths of time, and then the contents of the plant are analysed to see which are radioactive. This type of experiment showed the carbon from $^{14}CO_2$ in the atmosphere being built up into sugars, other carbohydrates, fats and proteins. For protein synthesis the plant also requires elements such as phosphorus and sulphur, obtained from the soil.

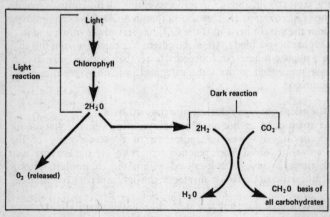

Figure 32. Summary of light and dark reactions

Heterotrophic nutrition

Every heterotrophic organism must acquire the organic substances it needs and render them into a suitable form before they can be absorbed into the body tissues and used in metabolism. In general several stages may be recognised in heterotrophic nutrition.

The process of taking in food is called **ingestion** and the way it is accomplished constitutes the animal's **feeding mechanism**. The food has usually been searched for and/or captured and for this a variety of methods and structures is employed. The ingested food is often in a complex insoluble form and it is converted into soluble compounds by the process of **digestion**. The soluble compounds are passed across living membranes to be made available to the cells; this process is known as **absorption**. After absorption the soluble compounds are distributed to the body cells, often by a circulatory system, and are built up into complex materials (**assimilation**) or broken down for energy release (**cell respiration**). The undigested, unabsorbed food materials are removed from the body by the process of **egestion**.

Egestion must not be confused with excretion. Excretory products are waste substances produced during metabolism and they have formed part of the cellular constituents; undigested food has never formed part of the cellular constituents but has only been in the cavity of the alimentary canal.

Feeding mechanisms

Within the animal kingdom many feeding mechanisms have evolved which deal with a great variety of food material, and heterotrophs can be classified according to the type of food they eat. **Herbivores** feed on plants, **carnivores** feed on animals, and **omnivores** feed on plants and animals. These terms are sometimes used only with reference to mammals, for example a carnivore would be defined as a flesh-eating mammal. In addition there are **liquid feeders** which consume a variety of plant and animal fluids, and **filter feeders** which live on small particles living in the water.

Herbivores must be able to crop the often tough vegetation and if maximum value is to be gained from the plant material the cells must be broken open and the cellulose digested and absorbed: only insects, mammals and molluscs (snails) have evolved the apparatus to do this. The protein in plants contains few essential amino acids so herbivores must consume large quantities of vegetation and they spend a great deal of the day feeding. Herbivores form the main prey of carnivores and they have evolved many devices for avoiding capture, for example deer rely on sustained speed to escape and live in herds for protection.

Carnivores eat whole animals or parts of them and to find and capture their prey a wide range of devices have evolved. These adaptations take various forms such as the great acceleration in speed, and dagger-like canine teeth for killing seen in the cats. One or more of their senses, such as sight or smell, is very highly developed. The food will contain a high proportion of essential amino acids so most carnivores feed at intervals.

Omnivores show a mixture of adaptations between carnivores and herbivores; they spend less time feeding than herbivores but more than carnivores. Feeding in man (an omnivore) will be dealt with later in the chapter, along with teeth.

Amoeba proteus is a fresh-water protozoan which lives in ponds. It feeds on microscopic organisms such as bacteria, algae and other protozoa living in the water. These organisms release chemicals into the water, the *Amoeba* senses them and moves towards the potential prey by the extension of pseudopodia. When an organism comes in contact with plasma membrane (plasmalemma) of the *Amoeba* it appears to 'stick' to it and a cup-shaped pseudopodium is put out around the prey which becomes enclosed in a drop of water to form a food vacuole; this is then taken into the cytoplasm for digestion and absorption.

Hydra is a coelenterate and is found in ponds, ditches and slow-running streams. It is a carnivore, feeding on small organisms such as *Daphnia* (water fleas). One species is *Chlorohydra viridissima* the green hydra. Its green appearance is caused by algae living in the endoderm cells. On the tentacles are batteries of stinging cells or **nematoblasts** (see page 24): when suitable prey brushes against the cnidocil or trigger of the nematoblasts they are discharged, the barbs pierce the prey first and then the everted thread is shot into it together with a paralysing fluid; other threads hold the prey. The tentacles bend over the mouth which opens to receive the food; digestion then occurs. Nematoblasts can only explode once; after use they are discarded and replaced from reserve (interstitial) cells.

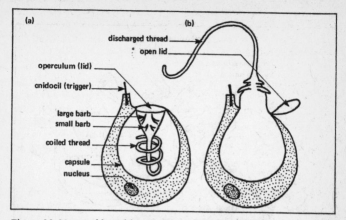

Figure 33. Nematoblast: (a) undischarged, (b) discharged

The **locust** is an insect; it is a herbivore and eats almost any kind of vegetation and its mouth parts are made of toughened cuticle and are well adapted for cropping and grinding the tough vegetation. The mouth parts consist basically of four parts arranged in layers one below the other. The **labrum** (upper lip) partially covers and protects the two large black **mandibles** (jaws), which have a sharp cutting edge and a raised grinding surface. They are rocked towards and away from each other (i.e. transversally) by powerful muscles and they cut and grind the vegetation. Below the mandibles are two **maxillae** (accessory jaws). They possess a cutting surface operated by muscles and they assist the mandibles in cutting the vegetation and pushing it into the mouth. When not in use they are enclosed in a sheath. Each maxilla has a long, jointed **palp** which is used to feel and test the food. The **labium** (lower lip) forms the floor of the mouth and it also possesses a pair of palps. Opening into the mouth is the duct carrying saliva from the salivary glands; the saliva helps to moisten the vegetation. In Africa and Asia locusts have been responsible for much human starvation. A medium-sized swarm will contain about 1,000,000,000 locusts each consuming its own weight in food per day; such a swarm would require 3,000 tonnes of plant food per day.

Mammals obtain food by using their teeth. The different arrangements of teeth found in herbivores, carnivores and omnivores are described later.

61

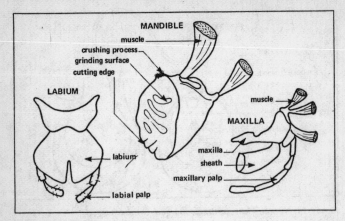

Figure 34. Mouthparts of a locust

Liquid feeders

Liquid feeders are highly specialised animals which can either extract concentrated solutions from body fluids of animals or plants, or break down solid food into a liquid form before ingestion.

Gut parasites such as the **tapeworm** absorb the digested food of their hosts directly across their outer surface.

The **housefly** feeds on organic matter such as faeces, compost and human food and it is owing to its indiscriminate feeding habits that it can transfer micro-organisms such as bacteria to human food. Many of the bacteria are harmless but others are responsible for such diseases as cholera, dysentery and salmonella (food poisoning). The feeding apparatus is a single hollow tube called a **proboscis** which is situated under the head. At the base of the proboscis are two fleshy lobes which are traversed by a system of food canals (pseudotracheae) which lead into the oesophagus into which the salivary duct from the salivary gland leads. When ready to feed the fly extends its proboscis and saliva is released down the salivary duct, through the pseudotracheae and on to the food. The saliva contains enzymes which begin to digest the food and the semi-digested liquid is then pumped up through the pseudo-tracheae and oesophagus and into the stomach by contraction of muscles in the upper part of the proboscis. The semi-digested fluid is often regurgitated and is deposited as a vomit spot on any surface where the fly lands.

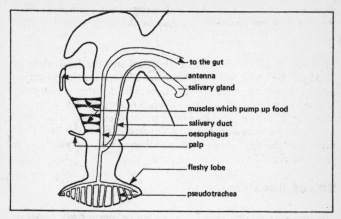

Figure 35. Section through the head and proboscis of a housefly

Male **mosquitoes** feed on plant juices by piercing into the phloem, but the female must feed on blood prior to egg laying. When the female is taking a blood meal, for example from man, the four **stylets** (formed from two mandibles and two maxillae) pierce the skin and when a capillary is reached saliva containing an anti-coagulant is released down the hypopharynx into the capillary; this prevents the blood clotting when it is sucked up by the tubular labrum. When the mouth parts are not in use they are enclosed in the rolled labium. Other liquid feeders are aphids (greenfly) and butterflies. **Aphids** are insects which feed on plant juices which they suck from leaves and stems. Their mouth parts are greatly elongated to form a piercing and sucking proboscis. The two maxillae form a tube which is pushed into plant tissues to reach the phloem from which nutrients are extracted. In **butterflies** the maxillae form a tubular proboscis through which nectar is sucked from flowers.

Filter feeders

Filter feeders are always aquatic and they feed on small organisms suspended in the water. A current of water is drawn towards the organism by cilia or by the movement of limbs. The food particles must be extracted from the water and this is achieved by trapping them in mucus and/or filtering them off through fine hair-like processes such as cilia or setae on the limbs. A rejection mechanism is usually present as not all particles are edible. Finally the selected particles must be moved towards the mouth and into the alimentary canal. As the suspended organisms are usually very small, vast quantities of water may be drawn towards the filter feeder.

In **mussels** cilia on the gills collect and sort the food while in *Daphnia* the limbs and their setae are adapted for this function. *Paramecium* is a protozoan which lives in fresh water; it is covered in cilia and it feeds when at rest or when moving slowly. A current of water is drawn into its oral groove where long cilia reject unsuitable particles and waft suitable particles down the 'gullet'; at the base of this the food particles are taken up into the cytoplasm in food vacuoles.

Site of digestion

The majority of organisms possess a gut or **alimentary canal** in the cavity of which the ingested food is digested. This is known as **extracellular digestion** as digestion occurs outside the cells. In contrast a few organisms have **intracellular digestion** where solid food particles are taken up into the cells and then digested in food vacuoles within the cells. *Amoeba* has intracellular digestion only (see page 60). *Hydra* has both types: extracellular digestion in the enteron reduces the food to small particles and these are taken in by pseudopodial cells of the endoderm lining the enteron, digestion being completed intracellularly within food vacuoles.

Some gut parasites, such as *Taenia* (tapeworm), do not possess a gut or any equivalent structure. The soluble end-products of digestion produced by its host are absorbed directly into its body across its surface.

Digestion in mammals

Digestion is carried out within the cavity of the alimentary canal. This is a very long tube which begins at the mouth and ends at the anus and along its length parts have become modified for specific functions; associated with the canal are various glands. The food is subjected to a combination of physical and chemical digestion.

Physical digestion is carried out by **teeth** and by rhythmical **muscular contractions** of the gut wall. Mammals possess four types of teeth, each adapted to carry out a specific function: in general the **incisors** bite off the food, the **canines** kill it, and the **premolars** and **molars** break it down prior to swallowing. Details of teeth are given at the end of the chapter. Muscular contraction is also responsible for moving the food along the gut. Circular muscles in the wall of the alimentary canal contract behind the food and force it along; this process is known as **peristalsis**.

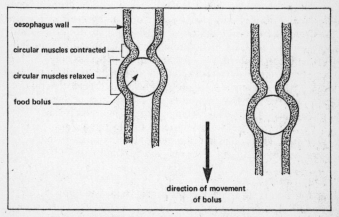

Figure 36. Diagram to show peristalsis in the oesophagus

Chemical digestion is carried out by **enzymes**. These are secreted from glands lying in the gut wall and from others lying outside the gut but connected to it by ducts. Enzymes break down large molecules into small molecules by the addition of water; this is known as **hydrolysis**.

Enzymes work at a particular pH, therefore changes in pH occur along the alimentary canal. There are many enzymes involved in the breakdown of food molecules but in general they fall into three groups: carbohydrases which act on carbohydrates, lipases which act on fats, and proteases which act on proteins. **Mucus** is secreted along the length of the alimentary canal to prevent proteases digesting the cells; the mucus also lubricates the gut and facilitates the passage of food along it.

Many experiments can be carried out on enzyme action, but they tend to demonstrate three basic principles: that enzymes only work on one substrate, and that enzymes work with maximum efficiency at a particular pH and temperature (the optimum pH and temperature).

Experiment: To demonstrate the effect of pH on enzyme action (using the enzyme pepsin and the protein, coagulated egg white, as the substrate).

Egg white is diluted with water, stirred, and the mixture is heated until a cloudy suspension is obtained. To test tube A 20 mm^3 egg white, 10 mm^3 1 per cent solution of pepsin, and six drops of bench hydrochloric acid is added; to test tube B 20 mm^3 egg white, 10 mm^3 1 per cent solution of pepsin, and six drops of sodium bicarbonate solution. In addition controls C, D and E, containing 20 mm^3 egg white and 1 per cent solution of pepsin only, 20 mm^3 egg white and six drops of sodium bicarbonate solution only, and 20 mm^3 egg white and six drops of bench hydrochloric acid only, are also set up. The test tubes are placed in a beaker of water maintained at 37°C (body temperature). Digestion is said to have occurred when the cloudy suspension is lost and a clear solution produced. This occurs in test tube A only; this indicates that pepsin digests the protein egg white only in acid conditions.

Experiment: To determine the effect of temperature on enzyme action (using the enzyme ptyalin in saliva and the substrate, starch).

A starch solution is made up and tested with iodine solution to obtain the characteristic blue/black colour. Five test tubes each containing 50 mm^3 2 per cent starch solution are placed separately in beakers of water at 0°C, 20°C, 40°C, 60°C and 80°C, and left for five minutes. Rows of drops of iodine are placed on a white tile. The mouth is rinsed out and a sample of saliva is collected, an equal volume of water is added, the solution mixed and 10 mm^3 samples are added to each tube. **As soon as** the contents of each test tube are mixed, one drop is withdrawn and tested for starch. This procedure is repeated at sixty-second intervals. When a sample gives a negative result with iodine solution it can be assumed that the contents of the test tube have been digested, i.e. starch has been digested by the enzyme ptyalin. Digestion occurs most rapidly at 40°C and least rapidly (if at all) at 0°C and 80°C. The optimum temperature for ptyalin action is 40°C (body temperature).

Feeding and digestion in man

Man requires a diet containing carbohydrates, fats, proteins, vitamins, minerals, water and roughage, but for a diet to be balanced these components must be present in the correct proportions. The food intake must provide substances which can be used in respiration to produce energy, it must enable growth and cell replacement to occur, and it must provide substances for certain secretions and for metabolic processes.

The amount of **energy** contained in food is measured in **joules** and **kilojoules** (or calories and kilocalories). A joule is a unit of energy and 1,000 joules = 1 kilojoule (kJ). 4·2 joules = 1 calorie. A calorie is also a unit of energy and it can be defined as the amount of heat needed to raise the temperature of 1 g of water by 1°C. Calories are in less common use now and joules and kilojoules should be used in preference. The main components which provide energy are carbohydrates and fats; 1 g carbohydrate provides 17 kJ and 1 g fat 39 kJ. The number of kJ required per day depends on the size, age, occupation, sex and activity of the person. An indication can be obtained by comparing energy requirements for different people, e.g. a heavy manual worker requires about 24,000 kJ, an office worker about 11,000 kJ, a child of eight about 8,500 kJ, and a one-year-old baby about 4,200 kJ.

In addition **carbohydrates** can be stored as **glycogen** in liver and muscles. **Fats** also form food stores and deposits under the skin are important in insulation. Fats are also important for the formation of cell membranes.

Proteins are required for growth and cell replacement, and for the production of substances such as enzymes and hormones. During digestion the protein is broken down into its constituent **amino-acids** which are absorbed and in the body cells are reassembled to form whichever protein is required. The body cannot make amino-acids but it can convert some amino-acids into others; those produced in this way are known as **non-essential** amino-acids. **Essential** amino-acids cannot be produced by the body and must be provided in the diet. There are about ten or eleven of these and animal protein, such as meat, contains a high proportion of them. Excess amino-acids are deaminated in the liver (see pages 75 and 121) to produce carbohydrates; these can be used in respiration or converted to glycogen and stored. 1 g of protein yields 17 kJ.

Vitamins are complex chemical compounds also known as accessory factors. They are required in minute amounts but are essential for the normal chemical activities of the body. If one is lacking in the diet a **deficiency disease** will occur and it can usually be cured by including the appropriate vitamin in the diet. A full vitamin chart is given on page 227. The discovery and importance of vitamins was realised by Sir Frederick Gowland **Hopkins** in his classic experiment on rats.

A wide variety of simple inorganic **mineral salts** is essential for the construction of certain tissues and for certain chemical activities in the body. For example, **iron** is required for the formation of haemoglobin in red blood corpuscles, **iodine** is necessary for the production of the hormone thyroxine, and **calcium** and **phosphorus** are required for the construction of teeth and bones. (See page 227.)

Water makes up a large proportion of protoplasm, and as enzymes only work in solution it plays a very important part in all metabolic processes. It is also important for the transport of substances in solution and for temperature regulation.

Roughage consists mainly of cellulose from plant cell walls. It adds bulk to the food and stimulates peristalsis.

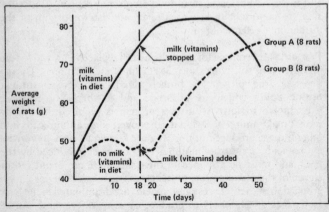

Figure 37. Results of Gowland Hopkins's experiment on rats

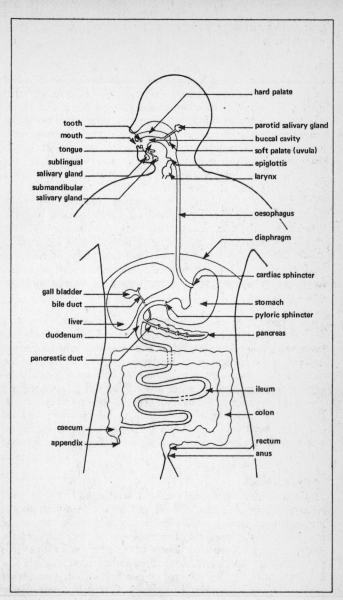

Figure 38. The human alimentary canal

69

Digestion in man

In the buccal cavity

The buccal cavity is situated between the mouth and oesophagus. The food may be bitten off by the incisors and canines and passed to the sides of the cavity where it is chewed by the premolars and molars. Chewing (mastication) breaks the food down; this allows it to be mixed with saliva, increases the surface area on which enzymes can act and makes it easier to swallow. Saliva is secreted from three pairs of salivary glands, the parotid, submandibular and sublingual, and enters the buccal cavity by salivary ducts. Its flow is initiated by the sight, smell, taste or expectation of food. Saliva contains the enzyme **ptyalin** or **salivary amylase**, mucus (mucin), sodium bicarbonate and sodium chloride, and water. Salivary amylase converts the polysaccharide, starch, into the disaccharide, maltose. Sodium bicarbonate and sodium chloride provide the correct pH (neutral or slightly alkaline) for amylase action. The mucus lubricates the food and makes it easier to swallow. When in solution the food can be tasted by the taste buds on the dorsal surface of the tongue; they are sensitive to sour, salt, bitter and sweet foods.

The tongue and cheek muscles help to mix the food particles with the saliva and then the particles are collected into a ball (bolus) prior to swallowing. During swallowing the tongue presses upwards and back against the hard palate forcing the bolus to the back of the buccal cavity (pharynx). At this point swallowing becomes an automatic or reflex action. The soft palate rises and closes the entrance to the nasal cavity; the opening to the larynx, the glottis, is closed by a flap of cartilage, the epiglottis; the food bolus is forced into and down the oesophagus or gullet, by **peristalsis**, and it enters the stomach through the cardiac sphincter.

In the stomach

The stomach is a muscular sac with flexible walls. It can store relatively large amounts of food and allows mammals to eat intermittently. Food is retained by closure of the pyloric sphincter at the end of the stomach. Depending on the time required for gastric digestion, food may remain in the stomach for up to five hours. In the stomach any fat is melted and the food is acted upon by gastric juice which is secreted by gastric glands in the stomach wall. Its flow is initiated by the sight, smell, taste or expectation of food, but is maintained by the hormone **gastrin** (see page 148).

Gastric juice contains the enzymes pepsin and rennin. **Pepsin** acts on proteins and converts them into peptones; it is secreted in an inactive form called pepsinogen to prevent digestion of the cells in which it is produced, and it is converted to pepsin in the cavity of the stomach by hydrochloric acid. **Rennin** is very important in babies: it clots or coagulates milk by converting the soluble milk protein, caseinogen, into insoluble casein so it may be retained in the stomach and be acted upon by pepsin. Hydrochloric acid present in gastric juice provides the optimum pH, of about $2 \cdot 0$, for the gastric enzymes; the acid also activates pepsinogen, kills many bacteria that have entered the stomach with the food, and stops the action of ptyalin. Mucus is also present in gastric juice; it lubricates the food and protects the stomach wall from enzyme action.

Rhythmic peristaltic movements of the stomach wall reduce the stomach contents to a semi-fluid called **chyme**. When digestion in the stomach is complete the pyloric sphincter relaxes at intervals and the chyme is passed a little at a time into the first part of the small intestine, the duodenum.

In the small intestine (duodenum and ileum)
In the duodenum the food is mixed with three secretions, bile, pancreatic juice and intestinal juice. **Bile** is a green alkaline fluid which does not contain enzymes; it is produced in the liver, stored in the gall bladder and released into the duodenum down the bile duct. Bile salts emulsify fats by lowering their surface tension and converting them into small droplets; this provides a large surface area on which the fat enzyme lipase can act. **Pancreatic juice** is produced in the pancreas and enters the duodenum down the pancreatic duct. Acid food passing into the duodenum causes cells in its walls to produce the hormone **secretin**; this is passed into the bloodstream and on reaching the pancreas causes the release of pancreatic juice. The three main pancreatic enzymes are: pancreatic **amylase** which breaks down starch to maltose, **lipase** which breaks down emulsified fats into fatty acids and glycerol, and **trypsin** which breaks down proteins into peptides. Trypsin is secreted in an inactive form called trypsinogen and it is converted into active trypsin by the enzyme enterokinase which is produced in the walls of the small intestine.

Bile and pancreatic juice contain sodium chloride and sodium bicarbonate; these help to neutralise the acid chyme from the stomach to produce an alkaline **chyle** and thus provide the correct pH for the action of the pancreatic and intestinal enzymes.

Secretory cells in the wall of the small intestine (duodenum and possibly the ileum) produce an intestinal juice which completes digestion. These enzymes include:

maltase which breaks down maltose into glucose;

lactase which breaks down lactose into glucose and galactose;

sucrase which breaks down sucrose into glucose and fructose;

erepsin (peptidases) which converts polypeptides to amino-acids. Some lipase may be produced in the small intestine but the pancreas is the main source.

In the large intestine (colon and rectum)

The colon does not secrete enzymes. The material entering the colon consists of undigested material (mainly cellulose), bacteria, mucus and water. Much of the water is absorbed from the food and the remaining semi-solid faeces is passed into the rectum for temporary storage and after twelve to twenty-four hours is expelled at intervals through the anus.

Absorption

Some water, glucose, alcohol and drugs may be absorbed through the stomach wall into the bloodstream, but the main region for absorption is the ileum.

The ileum is well adapted for absorption. It has a very large surface area because of its considerable length (approximately 22 feet), its folded wall, and its inner surface which consists of numerous finger-like projections called **villi**. The surface area is further increased by the tiny microscopic extensions of the individual epithelial cells covering the villi; these are called **microvilli**. Within each villus is a dense network of blood capillaries which receive oxygenated blood from branches of the mesenteric artery and which distribute deoxygenated blood to the **hepatic portal vein**. In the centre of each villus is a **lacteal** derived from a lymph vessel and which forms part of the lymphatic system.

The end products of carbohydrate digestion (glucose, fructose and galactose) and protein digestion (amino-acids) pass through the epithelium and into the blood capillaries. This occurs by a combination of diffusion and active transport. Active transport requires the expenditure of energy, and chemicals known to inhibit respiration, and thus prevent energy release, prevent the uptake of these substances. These absorbed substances are then transported in the hepatic portal vein to the liver.

Fats may be absorbed as fatty acids and glycerol or as very small drops of emulsified fats. Some two-thirds of the fat enters the lacteals as emulsified fats. The lacteals unite to form lymph vessels which run into the thoracic duct; this empties its contents into the bloodstream at the junction of the left subclavian and left jugular veins in the neck. Fatty acids and glycerol are passed into the blood capillaries and taken in the hepatic portal vein to the liver.

Mineral salts, vitamins and some water are absorbed in the small intestine. Most of the water is absorbed in the colon, which forms part of the large intestine.

Bacteria living in the cavity of the intestines produce vitamins, notably vitamin K and vitamin B_{12}, during their metabolism. These can be absorbed together with the vitamins taken in as part of the diet.

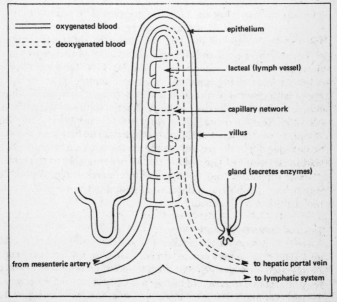

Figure 39. A villus (L.S.)

Experiment: To demonstrate the gut wall as a barrier (using visking tubing).

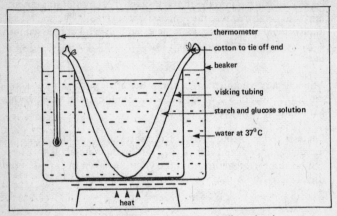

Figure 40. Experiment to demonstrate the gut wall as a barrier

The experiment is set up as in the diagram. The visking tubing represents the gut wall, the starch and glucose solution represent food in the cavity of the gut, the water in the small beaker represents the blood, and the wall of the large beaker represents the body wall. The experiment is kept at 37°C (body temperature). As soon as the apparatus is set up a sample of water from the small beaker is withdrawn and tested for starch with iodine solution and for glucose with Benedict's solution (Key Facts Revision Section, page 228). After twenty minutes a further sample is withdrawn and the tests repeated. The first tests should be negative but in the second sample the Benedict's test for glucose should be positive. The visking tubing is permeable to the small glucose molecules and not to the large starch molecules. This experiment demonstrates the necessity for digestion as only small molecules can be absorbed.

Fate of absorbed food

After absorption glucose is taken to the liver in the hepatic portal vein. The glucose may be used for respiration in the liver or it may be released into the general circulation to maintain the correct level; the concentration of glucose in the blood is usually between 90 and 100 mg/1000 mm^3 blood. In addition the glucose may be converted to glycogen under the influence of the hormone **insulin** which is secreted in the pancreas and carried to the liver in the blood. When the level of glucose in the blood falls below 80 mg/1000 mm³ blood the glycogen is converted back to glucose under the influence of hormones, e.g. adrenalin, and released into the circulation. Some glucose is converted to glycogen in the muscles. Only a limited amount of glycogen can be held by the liver and muscles; excess glucose not converted to glycogen is converted to fat and stored.

Fat, whether it is absorbed as fatty acids and glycerol or emulsified fat, enters the bloodstream. The fat may be used in respiration or for the production of cell membranes, or it may be stored. Fat is stored under the skin and around the kidneys and intestines and there is no limit to the amount that can be stored; when required for respiration it is taken to the liver.

After absorption amino-acids are taken to the liver in the hepatic portal vein and then released into the general circulation and to the cells which select the amino-acids they require for protein synthesis. Amino-acids and protein cannot be stored in the body as such and excess or unwanted amino-acids are taken to the liver and **deaminated**, i.e. the nitrogen or amine (NH_2) group is removed and converted into urea which is taken to the kidneys for excretion. The residue consisting of carbon, hydrogen and oxygen may be used as a carbohydrate and used in respiration or converted to glycogen or fat.

For a summary of liver structure and function, see page 228–9.

Alimentary canal in other mammals

Herbivores tend to have a proportionally longer gut than carnivores or omnivores as vegetation is more difficult to digest. To derive maximum value from plant food the complex carbohydrate cellulose from the plant cell walls must be digested. This requires the enzyme **cellulase** which is produced by very few organisms and is not produced by any mammal. In herbivorous mammals the enzyme is produced by micro-organisms (mainly bacteria and protozoa) which are housed in a specialised part of the alimentary canal; the cellulose is digested by them and the soluble end-products produced are absorbed by the mammal. In certain **ruminants** such as the cow, the micro-organisms are housed in three large sacs (extra 'stomachs'), the rumen, reticulum and omasum, situated at the base of the oesophagus and before the true stomach. After chewing, the food passes into the rumen where the majority of the micro-organisms are, and cellulose digestion begins. The food may be returned to the buccal cavity for further chewing ('chewing the cud'). Further cellulose digestion occurs in the rumen and reticulum through the walls of which the soluble products are absorbed. In the omasum water is absorbed and the food passes into the true stomach, the abomasum.

In **rabbits** the micro-organisms are situated in the **caecum** and **appendix**. The caecum is a large sac situated at the junction of the small and large intestines and it ends in a prominent appendix. In addition rabbits and shrews pass the food twice through the gut. On its first passage the food is only partially digested and the soft gelatinous faeces produced are eaten; during its second passage digestion is completed and the typical faecal pellets are then produced. This is known as **coprophagy** (refection).

The micro-organisms and the herbivores benefit from this arrangement. The micro-organisms are provided with a habitat and a constant food source, and any waste products are removed. The herbivore relies on the micro-organisms to digest the cellulose and then it absorbs the soluble products. Such a relationship where two species live in close association with each other and where both benefit from the arrangement is known as **symbiosis**.

Teeth

Teeth are derived from the skin covering the jaws and are set into sockets along the jaw bones. Characteristically mammals possess four types of teeth, each adapted to carry out a specific function: they are incisors, canines, premolars and molars.

Incisors are usually chisel-shaped and can be used for cutting, gnawing, tearing, pulling and defence. **Canines** are generally cone-shaped and are used for killing, tearing and fighting. Cheek teeth (**premolars** and **molars**) have a large surface area and more than one root for firm attachment in the jaw. They are used for grinding, crushing or slicing. The number, arrangement, structure and function of the different types of teeth is different in herbivores, carnivores and omnivores.

Structure of teeth Each tooth consists of a crown, neck and root. The crown is covered with enamel; it is the hardest substance in the body, is non-living, and consists of 96 per cent mineral salts

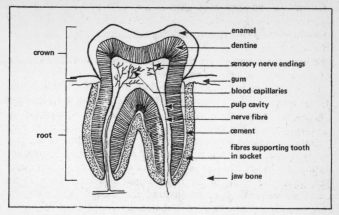

Figure 41. Vertical section through a human molar

(mostly calcium phosphate). Beneath the enamel and continuing into the root is a layer of dentine which forms the bulk of the tooth; it is a living layer which is similar to bone, and it is softer than enamel and contains fewer mineral salts. In the centre of each tooth is a pulp cavity containing nerve fibres, sensory receptors and blood vessels which supply food and oxygen to the dentine. These enter through an aperture at the base of the root. Cement holds the tooth in the socket and the fibrous membrane allows movement of the tooth in the socket.

Sets of teeth Normally mammals possess two sets of teeth. The first set is the **milk** or **deciduous** dentition; this is present in young mammals and consists of incisors, canines and premolars. The milk dentition is replaced by the **permanent** dentition which consists of incisors, canines, premolars and, in addition, molars.

Dental formula The number and arrangement of teeth in the skull (the dentition) varies in different species of mammal, so a convenient abbreviated method of showing the dentition is by the use of a dental formula. Symbols are used for the four types of teeth: i, c, pm, and m for the incisors, canines, premolars and molars respectively. For half the jaw only and against each appropriate symbol the number of each type of tooth in the upper jaw is put over the number of each type of tooth in the lower jaw.

Dental formula for man is i$\frac{2}{2}$ c$\frac{1}{1}$ pm$\frac{2}{2}$ m$\frac{3}{3}$ = 16 for half jaw.

Dentition of a carnivore (dog)

Dental formula $i\frac{3}{3}$ $c\frac{1}{1}$ $pm\frac{4}{4}$ $m\frac{2}{3}$

Carnivorous mammals are flesh eaters and must be able to kill, tear up and slice the prey prior to swallowing it. In the front of the mouth are small peg-like incisors which are used for nibbling meat from bones and for grooming. Behind the incisors are very large fang-like canines; these are used to kill the prey by piercing the jugular vein in the neck. The premolars and molars have their crowns elongated into ridges running along the line of the jaw. These ridges act like scissor blades and slice flesh from the prey and crunch the bones. The last upper premolar and the first lower molar are especially enlarged for this purpose and are known as the **carnassial teeth**; they are situated near the jaw joint where the greatest force can be exerted. The well developed temporal muscle and the low point of articulation of the lower jaw with the skull gives the jaw strength and excellent leverage. In addition only an up and down movement of the lower jaw is possible and again this adds to the strength of the jaw. Carnivores slice meat and then swallow it and there is very little chewing. The teeth do not have the same amount of wear as those of herbivores so do not have open roots and do not grow continuously.

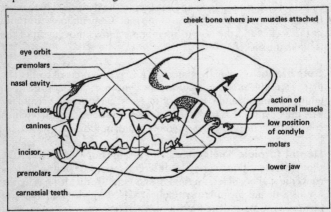

Figure 42. The skull and teeth of a dog.

Dentition of a herbivore (sheep)

Dental formula $i\frac{0}{3}$ $c\frac{0}{1}$ $pm\frac{3}{3}$ $m\frac{3}{3}$

The sheep must be able to crop and grind the tough vegetation. The incisors in the lower jaw are large and chisel-shaped and they have thicker enamel on their anterior face and so maintain a permanent hard cutting edge. The canines are often absent in herbivores but the lower canines in sheep have the same structure and function as incisors. The upper incisors and canines are absent and are replaced by a horny pad; the lower incisors and canines act against the horny pad to crop the grass. Before the cheek teeth is a gap, the diastema; it separates the newly cropped grass from that being ground by the cheek teeth. The premolars and molars are large and the grinding surface is increased by the teeth being folded into a W or M shape. The surface of the tooth consists of a series of enamel ridges and dentine troughs; these are formed as the dentine is softer and wears away more quickly than the hard enamel. The M-shaped ridges of the lower jaw fit into the W-shaped ridges of the upper jaw and they grind across each other to grind down the grass; this is aided by the circular jaw action. The jaw joint is high and this gives the well developed masseter muscle greater leverage. The teeth have persistently open roots and continue growing throughout life; this is essential as they are in constant use and would otherwise wear away.

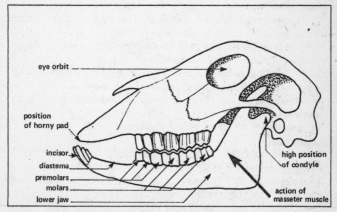

Figure 43. Skull and teeth of a sheep

Saprophytic nutrition

This method of nutrition is characteristic of many fungi and bacteria. It involves the absorption of soluble organic materials from dead or decaying matter. Enzymes are secreted by the organism into the organic material and extra-cellular or external digestion takes place. The soluble products then diffuse into the organism. Each saprophyte can only produce a certain range of enzymes and so tends to feed off a particular type of organic material. For example the fungi feeding on decaying wood are different from those feeding on dung.

The action of saprophytes is most important since it results in the gradual breakdown of plant and animal remains. This means that important elements are released back into the soil and may be reused by green plants (see nitrogen cycle, page 209). Saprophytic bacteria and fungi are also inconvenient to man since they cause spoilage of food products.

Key terms

Absorption Uptake of substances usually in solution across living cell membranes.

Autotrophic nutrition In which organisms manufacture their own food from inorganic material.

Digestion Process whereby complex insoluble substances are broken down into simple soluble substances which can then be absorbed.

Egestion The removal of undigested food materials from the body.

Enzymes Proteins which speed up the rate of chemical reactions but do not appear in or alter the end-products.

Heterotrophic nutrition In which organisms require an organic food supply from their environment.

Holophytic nutrition Involves building up food from inorganic substances using light energy trapped by chlorophyll.

Holozoic nutrition Taking in complex foods and digesting them internally.

Metabolism All the chemical processes that occur in living organisms.

Parasitic nutrition Food is obtained by living in or on another living organism, the host, which suffers some harm as a result.

Photosynthesis The building up of sugars from carbon dioxide and water using light energy trapped by chlorophyll.

Saprophytic nutrition Dead organic substances are digested externally and the products absorbed.

Chapter 3
Respiration: Gaseous Exchange

Respiration occurs in every living cell, and it is the way in which the cell obtains usable energy. The process may be represented by the equation

$$C_6H_{12}O_6 + 6O_2 \longrightarrow 6H_2O + 6CO_2 + 2880 \text{ kJ}$$

sugar   oxygen   water   carbon dioxide   energy

This is a breakdown or catabolic process in which oxygen is usually but not always involved. It is one of the most important parts of a cell's metabolism since life is only maintained at the expense of energy. Energy is needed to allow an organism to carry out work, such as muscular contraction, to conduct nerve impulses, to drive building up processes, such as protein synthesis, and to maintain body temperature in some organisms. The energy in the sugar molecule is in the form of chemical energy and is not directly available to the organism. Respiration converts this energy into chemical energy in an 'energy-rich' molecule called **adenosine triphosphate** (A.T.P.). A.T.P. occurs in every type of organism and directly supplies energy to any process requiring it. The energy from the sugar molecule is released slowly in a large number of steps, resulting in the production of A.T.P. molecules. Each step has its own specific enzyme. As in all processes involving energy changes, some is released in the form of heat. This heat has little value to most organisms, and so represents wastage. Homoiotherms, or warm-blooded animals, however, have evolved ways in which this heat may be trapped and used to maintain a constant body temperature. It is important that energy is immediately available within a cell. A.T.P. can satisfy this demand, because unlike a sugar molecule, it is quickly and easily broken down. A.T.P. is also a small molecule which can move about the cell more easily than a large sugar molecule. Each cell meets its own needs for energy, the final stages of respiration occurring within the mitochondria.

Respiration has come to involve a number of misleading terms. Generally the breakdown of sugar and the formation of A.T.P. is called **internal**, cell, or tissue **respiration**. The way in which the organism carries out the gas exchange involved in respiration and the transport of these gases to and from the cells is called **external respiration**. Breathing refers to the way in which higher animals ventilate the respiratory surface by body movements. Lower animals and plants, therefore, respire but do not breathe. The main features of internal respiration are the same in all organisms but external respiration is highly variable.

Internal respiration is of two types: **aerobic** respiration, in which oxygen is involved, and **anaerobic** respiration which does not require oxygen. The vast majority of organisms carry out aerobic respiration; a few organisms can exist permanently in the absence of oxygen (e.g. some soil-inhabiting bacteria), and other organisms can carry out anaerobic respiration for very short periods of time only (e.g. man). The first stage of internal respiration does not require oxygen but only releases a relatively small amount of energy. This stage involves splitting the stable six-carbon sugar molecule glucose ($C_6H_{12}O_6$) into two three-carbon molecules called pyruvic acid. Next the pyruvic acid is broken down, using oxygen, to release a large quantity of energy, carbon dioxide and water. This process is summarised in figure 44.

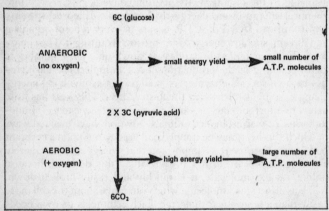

Figure 44. Internal respiration

Aerobic respiration leads to the complete breakdown of the sugar molecule, and the release of all its chemical energy. Anaerobic respiration only partly breaks down the sugar molecule. For this reason aerobic respiration is more efficient and provides the organism with more energy. The breakdown of pyruvic acid using oxygen is an example of the chemical reactions known as oxidations. The energy-holding molecule A.T.P. is used in a cycle within the cell. It is built up using the energy released from the breakdown of the sugar molecule (1), and then broken down during any process requiring energy. These processes are represented in the following equations.

(1) adenosine diphosphate (A.D.P.) + phosphate (P) + energy = A.T.P.

(2) A.T.P. $\longrightarrow$ A.D.P. + P + work

The final stages of anaerobic respiration, following on from the production of pyruvic acid, vary depending on which organism is concerned. In yeast, when no oxygen is available, pyruvic acid is converted into alcohol. Man uses this action of yeast in beer and wine making. The process may be represented by the equation

$$C_6H_{12}O_6 \longrightarrow 2C_2H_5OH + 2CO_2 + 210 \text{ kJ}$$
$$\text{glucose} \qquad \text{alcohol} \qquad \qquad \text{energy}$$

The production of alcohol is an example of **fermentation**. This term is used to describe anaerobic respiration in organisms which derive most of their energy from the process. Man not only uses yeast commercially but also other fungi and some bacteria to produce useful compounds from fermentation processes.

In animals, during strenuous exercise, oxygen may not be reaching the active cells quickly enough and this results in the cells having to carry out anaerobic respiration. This state of affairs will only exist for a short period of time since the accumulation of the products of anaerobic respiration causes fatigue. During the period of anaerobic respiration the animal develops an oxygen debt which is equal to the oxygen required to get rid of the products of anaerobic respiration. This is shown in figure 45.

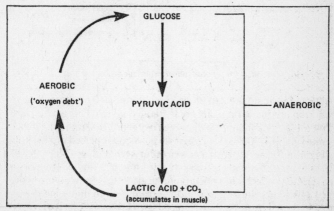

Figure 45. Anaerobic respiration in muscle

The 'oxygen debt' is the reason why it is necessary to go on breathing heavily for a short time after strenuous exercise.

Experiments on respiration It is possible to demonstrate by simple experiments the qualitative changes which occur during respiration. Respiratory activity, however, is often used as an indication of the general level of metabolic activity of a tissue or organism, so quantitative experiments are frequently carried out. These often involve measuring the changes in a gas volume and therefore allowances must always be made for any changes in temperature or atmospheric pressure which could alter the volume of a gas during the course of an experiment.

To demonstrate the products of fermentation by yeast

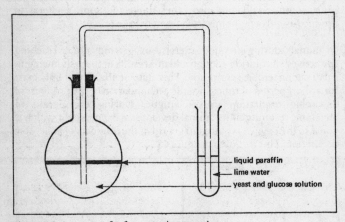

liquid paraffin
lime water
yeast and glucose solution

Figure 46. Apparatus for fermentation experiment

A 10 per cent suspension of dried yeast and a 10 per cent glucose solution are made up using cooled boiled water. This is to ensure that the water contains no dissolved oxygen. 5,000 mm³ of glucose solution and 1,000 mm³ of yeast suspension are placed in a flask, as shown in figure 46, and kept warm for several days. The layer of liquid paraffin prevents oxygen entering the liquid. A control experiment which does not contain yeast is also set up. During the first few hours the lime water turns from clear to milky indicating that carbon dioxide is being released. No carbon dioxide is obtained from the control experiment. After several days the glucose is distilled and alcohol, recognisable by its properties of odour and taste, is collected. No alcohol is obtained from the control experiment. The fact that alcohol still contains chemical energy, which has not been released during anaerobic respiration, may be demonstrated by igniting it. The alcohol burns releasing chemical energy in the form of heat and light.

To show the production of carbon dioxide during aerobic respiration

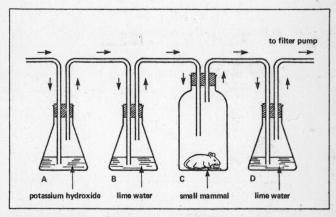

Figure 47. Apparatus to show the release of carbon dioxide

The apparatus shown in figure 47 may be used for a plant or small animal. When a plant is being used the bell-jar must be blacked out so that the plant is unable to carry out photosynthesis. In addition the flower-pot must be covered in polythene to prevent the respiration of any organism in the soil from influencing the result. The potassium hydroxide in A removes carbon dioxide (CO_2) from the air drawn into the apparatus. The lime water in B should remain clear throughout the experiment confirming that the air entering C is free of CO_2. The air then passes into C containing the organism and then into D which contains lime water to test again for the presence of CO_2. A filter pump draws air through the apparatus which is left for about half an hour or longer, depending on the organism being investigated. At the end of the experiment the lime water has become milky, indicating that the organism has released CO_2.

To demonstrate the release of energy during respiration

When a tissue respires rapidly much of the energy is released as heat. This may be demonstrated by using germinating seeds in which there is a rapid breakdown of sugars. The apparatus is set up as in figure 48.

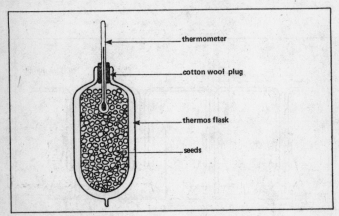

Figure 48. Energy release in seeds

The pea seeds are soaked in water for twenty-four hours and then left in a solution of sodium hypochlorite to kill off any micro-organisms on the seed coat. The seeds are rinsed in water before putting them in the thermos flask. A control experiment is set up, using seeds which have been killed by boiling. The temperature is noted at the beginning of the experiment, and then twice a day for several days. It is found that the temperature of the living seeds steadily rises, whilst that of the control fluctuates, reflecting changes in room temperature. This shows that the living seeds are releasing energy.

To demonstrate the uptake of oxygen during aerobic respiration

The apparatus used is shown in figure 49. A known volume of soaked seeds, which have been rinsed in hypochlorite solution to kill off any micro-organisms on the surface, is placed in the flask. Potassium hydroxide is placed in the central well to absorb CO_2.

The absorption is made more efficient by also putting in some filter paper, thus creating a greater surface area.

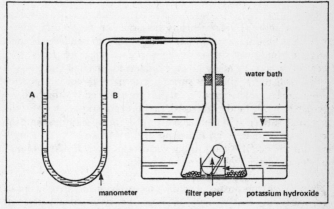

Figure 49. Uptake of oxygen in respiration

A control experiment is set up using the same volume of dead seeds. The apparatus is left in a water bath for five minutes, with the manometer disconnected, so that the air in the flasks is at atmospheric pressure and the temperature at that of the water bath. The manometer is then connected to the flask. Any CO_2 released will be absorbed by the potassium hydroxide. Therefore if any oxygen is absorbed by the seeds this will result in a decrease in the volume of air in the flask and the liquid in the arm B of the manometer will rise. Any other factors producing a change in the volume of air in the flask will be indicated by changes in the liquid levels of the control manometer. The water bath will minimise the effect of any temperature changes.

After several hours it is found that the liquid in the manometer arm B, connected to the flask containing the living seeds, has risen. The liquid in the manometer of the control experiment has only fluctuated slightly. This shows that the living seeds have absorbed oxygen.

Gaseous exchange in animals

In small organisms the surface area to volume ratio is large enough for diffusion of respiratory gases across the whole body surface to be adequate. This is the situation in protozoa, such as *Amoeba*, coelenterates such as *Hydra*, and annelids such as the earthworm. In *Amoeba* and *Hydra* the diffusion distances within the organisms are short and movement of respiratory gases through them is mainly by diffusion.

In larger, more active organisms the surface area to volume ratio is too small for diffusion across the body surface to be adequate and a special respiratory surface is developed which is restricted to a certain region. In many aquatic organisms external and internal gills form the respiratory surface; external gills are outgrowths from the body whereas internal gills are enclosed within the body where they are better protected. In most terrestrial organisms the external surface is impermeable to prevent desiccation, and the respiratory organ is an intucking of the surface and is enclosed within the body. Air-breathing vertebrates have developed lungs and insects have evolved a tracheal system.

Many organisms possess a ventilating mechanism whereby the air or water in contact with or near the respiratory surface is continuously being renewed. This is particularly important in aquatic organisms as the water will contain relatively little dissolved oxygen and diffusion of oxygen will be slow.

Respiratory surfaces have several features in common. They have a large surface area often achieved by branching or folding and this increases the amount of gases which can be exchanged; they are thin so diffusion is rapid as diffusion distances are short; they are moist as respiratory gases can only pass across respiratory surfaces in solution; they are permeable so respiratory gases can diffuse through them; they have a good blood supply (if the organism possesses a circulatory system) to carry the respiratory gases between the respiratory surface and body cells; it is important to maintain a steep diffusion gradient across the respiratory surface so the gases can always diffuse from a high to a low concentration.

Types of respiratory surface

In the **earthworm** the respiratory surface is the skin or epidermis. It is only one cell thick (therefore very thin) and it is kept moist by secretions from gland cells. It contains a vast network of looped capillaries which come very near to the surface. Oxygen from the air spaces in the soil dissolves in the surface moisture of the epidermis and diffuses into the blood capillaries. The blood contains haemoglobin in solution in the plasma with which the oxygen combines and this increases the oxygen-carrying capacity of the blood.

Insects possess a **tracheal system** which extends throughout the body. The exoskeleton is impermeable but it is perforated on either side of the thorax and abdomen by ten holes or **spiracles**, the entrances of which are protected by valves to reduce water loss. The spiracles lead into an extensive system of narrow tubes called **tracheae** which are supported by spirals of chitin to keep them open. They end in microscopic hair-like **tracheoles** which are $1 \mu m$ in diameter; they contain fluid and they run between all the cells of the body.

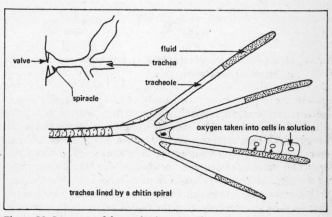

Figure 50. Structure of the tracheal system of insects

Oxygen passes by diffusion through the spiracles, along the tracheae and into the tracheoles where it dissolves in the fluid. The cells adjacent to the tracheoles produce waste products and they have a higher osmotic pressure than the fluid in the tracheoles; this causes fluid containing oxygen to be withdrawn from the tracheoles and into the cells. The oxygen enables the waste products to be converted into other substances, the osmotic pressure in the cells falls, and the fluid, now containing carbon

89

dioxide, moves back into the tracheoles. Ventilation of the tracheal system is brought about by movement of the thorax and abdomen and by air sacs along the tracheae acting like bellows. This method of gaseous exchange is very efficient as oxygen can be passed directly to the cells.

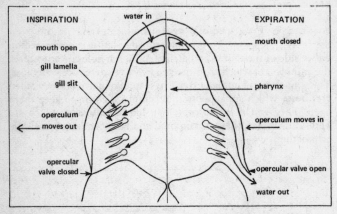

Figure 51. A section through the head of a fish showing the breathing mechanism

In **fish** the respiratory surfaces consist of four pairs of **gills** which lie between five pairs of gill slits and they are covered by a muscular operculum. The slits are continuous with the buccal cavity and pharynx and with the surrounding water via the opercular valve. To breathe the fish opens its mouth, the opercular valve is closed and the floor of the buccal cavity and pharynx is lowered by muscle action: this increases their volume and decreases the pressure and water is drawn in. The operculum is moved out and the floor of the buccal cavity and pharynx is raised and the water moves through the gill slits and across the gill lamellae where gaseous exchange occurs. The water is expelled through the opercular valve by the inward movement of the operculum with the buccal cavity and pharynx still in a raised position. The fish must have a continuous supply of water passing across the gills. To maintain a steep concentration gradient and to ensure maximum gaseous exchange the blood and water flow in opposite directions (**counterflow system**). As blood flows across the respiratory surface it meets water which has had less and less oxygen extracted from it.

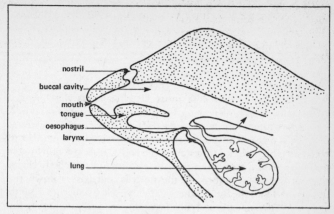

Figure 52. The respiratory organs of a frog

Frogs exhibit several methods of gaseous exchange during their life history. Before hatching from the spawn the embryo obtains oxygen and gets rid of carbon dioxide by diffusion through the whole body surface. On hatching, three pairs of frilly external gills develop just behind the head; they have very thin walls and a large surface area and blood is pumped through them. After about six days the external gills are absorbed and replaced by internal gills which are very similar in structure and functioning to those of fish.

Adult frogs carry out gaseous exchange by three methods. In water nearly all the gas exchange occurs through the **skin** and on land nearly half. The skin is very thin, it is kept moist by the water in its environment or by secretions from gland cells in the skin and it is richly supplied with blood capillaries. In addition gas exchange can occur within the **buccal cavity**: the mouth is closed, the nostrils opened, and the floor of the buccal cavity is lowered by muscular action; this increases its volume and thus reduces its pressure and air enters the buccal cavity. The lining of the buccal cavity is thin and moist and has a rich supply of blood capillaries into which the oxygen diffuses.

The **lungs** are only used to any extent when the frog is very active on land. They are ventilated by a modification of the method of gaseous exchange in the buccal cavity. Air is drawn into the buccal cavity in the same way and then the nostrils close, the floor of the buccal cavity is raised and air is forced through the larynx into the lungs where gaseous exchange occurs. The nostrils then open and the lungs empty due to their own elasticity and the pressure on them from other body organs. The frog does not possess ribs, diaphragm or intercostal muscles.

In **birds** gaseous exchange must be efficient as they need a continuous supply of oxygen because they are homoiothermic, very active, and often need to fly considerable distances without rest. Birds possess **lungs** but these are connected to a number of **air sacs** which extend throughout the body. Muscle action lowers the sternum and moves out the ribs, this enlarges the body cavity and air is drawn in through the nostrils and passes down the trachea and into the lungs where gaseous exchange occurs. The air passes into the air sacs, and then breathing out occurs whereby the air is passed from the air sacs and through the lungs again where further gaseous exchange can occur. This process is very efficient as the air passes through the lungs twice and all the air in the lungs is changed during one breath due to the action of the air sacs. During flight the breathing mechanism is reinforced by the action of the flight muscles alternately raising and lowering the sternum. Oxygen is carried as oxyhaemoglobin in oval red blood cells which possess a nucleus.

Gaseous exchange in man

It is essential for any mammal to have an efficient respiratory system as they are active homoiothermic organisms and have a high metabolic rate.

The **respiratory system** consists of air passages and lungs and these are situated in the head, neck and thorax. The thoracic cavity forms the upper part of the trunk and it is bound by the sternum at the front, the ribs and intercostal muscles at the side and the thoracic vertebrae at the back. The floor of the cavity is formed by the diaphragm, a muscular and fibrous sheet of tissue which extends across the body cavity and separates the thorax from the abdomen. Man possesses two thin-walled, spongy, elastic lungs which are situated on either side of the heart in the thorax. They are enclosed in very thin pleural membranes which enclose a pleural cavity. The membranes secrete pleural fluid which lubricates the surfaces in contact with the lungs and walls of the thorax and thus allows friction-free movement during breathing. Air is drawn into the lungs through the nostrils and it then passes into the nasal cavity. This cavity is richly supplied with blood capillaries and is lined by ciliated epithelium (see page 15); between the ciliated cells are mucus-secreting cells. Here the air is warmed, filtered and moistened before passing through the pharynx, glottis and larynx and into the trachea. During swallowing the glottis is covered by a flap of cartilage, the epiglottis. The walls of the trachea are supported by C-shaped cartilages which prevent the trachea from collapsing when the air pressure is reduced during inhalation. The cells lining the trachea possess cilia and between the ciliated cells are mucus-secreting cells; the mucus traps micro-organisms and dust in the inhaled air and the beating cilia move it up the trachea to the pharynx when it is then swallowed as phlegm. The trachea divides into two bronchi, which are also supported by cartilages, and a bronchus goes to each lung. Within each lung the bronchus passes into bronchioles which branch repeatedly and end in minute, pouch-like, thin-walled air sacs or alveoli: they are covered in blood capillaries where gaseous exchange occurs. Exhaled air moves in the opposite way.

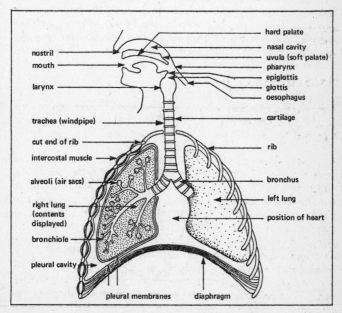

Figure 53. Respiratory structures in the head and thorax of man

Exchange of gases in the lungs

This occurs in the thin-walled alveoli which are in very close association with an extensive network of blood capillaries. There are about 700 million alveoli and they are lined with a thin layer of fluid. The capillaries receive deoxygenated blood from the pulmonary artery, and distribute oxygenated blood to the pulmonary vein which leads to the heart.

After inhalation there will be a high concentration of oxygen in the air in the alveoli, and a low concentration of oxygen in the blood capillaries as the deoxygenated blood has returned to the lungs from the body. The oxygen dissolves in the fluid lining the alveoli and diffuses through the very thin alveolar and capillary walls into the blood. Here it combines with the haemoglobin in the red blood cells to form oxyhaemoglobin.

Deoxygenated blood returning to the lungs in the pulmonary artery will contain a high proportion of carbon dioxide produced as a result of tissue respiration. It is carried mainly as bicarbonate ions dissolved in the plasma, but some is carried as carbamino-haemoglobin in the red blood cells. In the lungs the carbon dioxide is released from the bicarbonate ions by the enzyme carbonic anhydrase. As there is a high concentration of carbon dioxide in the blood and a low concentration in the alveolar air, it diffuses into the alveoli and is exhaled.

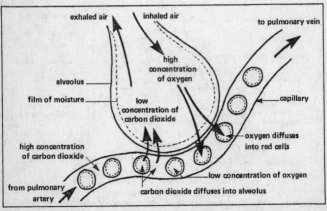

Figure 54. Gaseous exchange between an alveolus and capillary

Ventilation of the lungs

To exchange air in the lungs it is necessary to alter the volume of the thorax which will alter the air pressure in the lungs and respiratory passages. Any change in the volume of the thorax is transmitted to the lungs which are too thin to oppose it. At **inspiration** the volume of the thorax is increased and this is brought about by two movements: the intercostal muscles contract and the rib cage swings up and out; the muscles of the diaphragm contract and it is pulled down from its normal domed position. This increases the volume of the thorax, the air pressure is thus reduced and air enters through the nostrils and passes down the air passages to the lungs.

Expiration is partly a passive process. The volume of the thorax is reduced and this is brought about by the following. The intercostal muscles relax and the rib cage moves down under its own weight; the diaphragm muscles relax and this together with pressure from the organs and muscles of the abdomen causes it to move up into its original domed position. The elasticity of the lung tissue causes the lungs to move inwards, thus the volume of air in the thorax is reduced, the air pressure is increased and air is expelled.

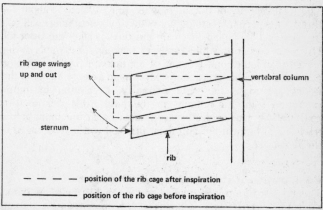

Figure 55. *The upward and outward movement of the rib cage at inspiration*

Diffusion gradient

It is important to maintain a steep diffusion gradient across the respiratory surface and this is achieved by a combination of factors. The lungs are well ventilated ensuring fresh supplies of inhaled air; the diffusion distances are short; the oxygen once in the blood is combined with haemoglobin and thus removed from solution; and lastly the pulmonary vein carries oxygenated blood away from the lungs quickly and deoxygenated blood is continually passing to the lungs in the pulmonary artery.

Control of breathing rate

The breathing rate of a normal adult is between twelve and twenty breaths per minute. In the medulla oblongata of the brain are groups of nerve cells which form the **respiratory centre** and it is connected by nerves to the diaphragm and intercostal muscles. The main factor controlling the breathing rate is the amount of carbon dioxide, and to a lesser extent, the amount of oxygen, in the blood. An excess of carbon dioxide in the blood stimulates the respiratory centre which sends impulses down the nerves to the diaphragm and intercostal muscles and the breathing rate is increased. The breathing rate decreases when the amount of carbon dioxide in the blood decreases. This is another example of **homeostasis**.

Capacity of the lungs

The total capacity of the lungs when fully inflated is about 5·5 litres, but a person breathing normally (quiet breathing) will take in and expel about 0·5 litres air and this is called the **tidal air**. However, deeper breathing can be carried out, for example during exercise, and the total amount of air that can be expired after maximum inspiration is about 4 litres in an average man and this is referred to as the **vital capacity**. Even after maximum expiration some air remains in the lungs; this is known as **residual air**. Due to dead space created by air remaining in the trachea and bronchi not all the air inspired will be exchanged in the alveoli.

Composition of atmospheric and exhaled air

The amount of nitrogen inhaled and exhaled is the same, but as the volume of oxygen absorbed is slightly more than the amount of carbon dioxide given out, so the percentage of nitrogen changes.

	Atmospheric Air per cent	Exhaled Air per cent
Oxygen	20·95	16·40
Carbon dioxide	0·04	4·10
Nitrogen	79·01	79·50
Water vapour	variable	saturated
Temperature	variable	body temperature

Table 1. Composition of atmospheric and exhaled air

To estimate quantitatively the percentage of oxygen and carbon dioxide in atmospheric and exhaled air

The apparatus for this experiment may vary but in all a known volume of air is held, usually in a graduated burette or capillary tube, and this is kept in a water bath throughout the experiment to minimise temperature fluctuations which would alter the volume of the gases. Sodium hydroxide is added to the atmospheric air sample to absorb the carbon dioxide; after five minutes the pressure of the air sample is adjusted to that of atmospheric air and the new volume of air (y) recorded. Alkaline pyragallol is then added to absorb the oxygen and after five minutes the pressure is again adjusted and the new volume of air (z) recorded. This procedure is repeated with a sample of exhaled air which is collected by immersing the burette or capillary tubing in water and breathing through the apparatus, thus displacing the water with exhaled air.

Calculation:

Original volume of air	$= x$ ml
Volume after carbon dioxide	$= y$ ml
Volume after oxygen absorbed	$= z$ ml
Volume of carbon dioxide in sample	$= x - y$
% of carbon dioxide in sample	$= \dfrac{x - y}{x} \times 100$
Volume of oxygen in the sample	$= y - z$
% of oxygen in the sample	$= \dfrac{y - z}{x} \times 100$

Gaseous exchange in plants

Plants have slower metabolic rates than animals and so do not need to obtain oxygen so rapidly. As a result no elaborate respiratory surfaces or circulatory systems have evolved in plants. Oxygen diffuses into the aerial parts of the plants through stomata or lenticels. The gas then circulates through a system of inter-cellular spaces and diffuses into cells through their moist walls. Young roots, which are not covered by a waterproof outer layer, obtain oxygen by diffusion anywhere along their surface. The oxygen is present in air-spaces in the soil, and for healthy root growth it is important to keep soils well aerated. Carbon dioxide produced by plant cells during respiration may not reach the atmosphere as it may be used up by photosynthesis. In the dark, carbon dioxide will diffuse out of the plant by the same route as the oxygen entered.

Key terms

Alveoli Air sacs of the lung surrounded by blood capillaries and in which gaseous exchange occurs.

Anabolic Building-up process.

Breathing Movements of the body associated with gaseous exchange.

Catabolic Breaking-down process.

Diffusion Movement of molecules of a gas or liquid from a region of high concentration to a region of low concentration.

Fermentation Anaerobic respiration in organisms which derive most of their energy from the process.

Lenticels Gaps in the bark for gaseous exchange.

Manometer Instrument which measures changes in pressure.

Respiration Process in which energy is produced in cells.

Aerobic: oxygen is used to break down sugar completely.

Anaerobic: sugar is only partly broken down in absence of O_2.

Respiratory surface Of animals, across which respiratory gases are exchanged with the environment.

Stomata Pores in epidermis of plants for gaseous exchange.

Thorax Upper part of the trunk from the neck to the diaphragm.

Chapter 4
Transport

Transport systems in animals

In large animals specialised surfaces located in one part of the body have evolved for the exchange of respiratory gases and for the absorption of soluble food substances. These substances must be carried to other parts of the body but the tissues are too bulky for diffusion alone to be adequate, so transport or circulatory systems have developed; these are referred to as **mass-flow systems**.

In organisms such as *Amoeba* and *Hydra* which have a small volume and therefore short diffusion distances, distribution of respiratory gases and soluble food through the organism by diffusion is probably adequate, but even at this level transport of substances is aided: in *Amoeba* by movement of cytoplasm and in *Hydra* by movement of the body and flagellar cells which keep the contents of the enteron circulating (see page 24). In many invertebrates and in all vertebrates a circulatory system has developed which consists basically of a system of blood vessels containing blood; this is circulated round the body, its flow being maintained by a muscular heart.

Blood in man

Blood can be regarded as a fluid tissue consisting of several types of cell suspended in a fluid called plasma; the cell constituents are red and white blood cells (or corpuscles) and platelets. The total volume of blood in the body of an average male is between four and five litres and in an average female about a litre less.

Plasma is a pale yellow alkaline fluid containing 90 per cent water but many compounds are suspended or dissolved in it. Important constituents of plasma are proteins such as fibrinogen and prothrombin which are important in blood clotting, globulins and antibodies. When fibrinogen is removed from it, plasma is known as **serum**. In addition the plasma will contain digested foods such as glucose, amino acids, fatty acids and glycerol, vitamins and salts, or more correctly the ions, of for example sodium, potassium, calcium, chloride, phosphate and bicarbonate. Nitrogenous waste

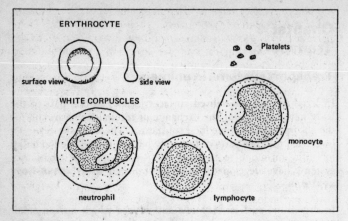

Figure 56. Components of blood (man)

products, such as urea, hormones, enzymes and small quantities of dissolved oxygen and carbon dioxide, will also be carried.

Red corpuscles or **erythrocytes** are minute biconcave discs with a diameter of $7 \cdot 5$ μm; they have no nucleus and consist of an elastic membrane enclosing spongy cytoplasm and there are about 5 million red cells per mm^3 of blood. In the cytoplasm is an iron-containing pigment called **haemoglobin**: it has an affinity for oxygen and combines with it in conditions of high oxygen concentration, for example at the lungs, to form oxyhaemoglobin. This compound is unstable and readily breaks down in the tissues where there is a low concentration of oxygen and the oxygen is thus made available to the cells. The biconcave shape of the corpuscle provides a large surface area for the absorption of oxygen. Red cells are made in the red bone marrow of short bones such as the sternum and vertebrae. They are said to have a life span of between 100 and 120 days; they are broken down in the liver and spleen and the iron is extracted and stored in the liver for future use. Red cells are made at the rate of 1 to $1\frac{1}{2}$ million per second.

White corpuscles or **leucocytes** differ from erythrocytes in that they have a nucleus, they are larger, varying in size between 8 and 15 μm, and they are fewer in number: there are about on average 8,000 per mm^3 so there is about one white cell to every six hundred red cells. Leucocytes are of different types and 70 per cent of them are made up from **polymorphs** (neutrophils or granulocytes); they have an irregular shape, granular cytoplasm and a two- or three-lobed nucleus. They are made in red bone marrow. Polymorphs are capable of independent movement by the formation of pseudopodia and they can pass through capillary walls into the tissues. They can engulf or ingest micro-organisms or dead cells and then digest them, a process very similar to feeding in *Amoeba*; this action is known as **phagocytosis** and thus polymorphs are said to be phagocytic. **Lymphocytes** make up 23 per cent of the white cells; they are smaller than polymorphs and the nucleus occupies most of the cell. They are produced in lymph glands, such as the tonsils and spleen, and in lymph nodes. They too are important in preventing infection and are responsible for the production of some antibodies. **Monocytes** are white cells with a slightly lobed nucleus; they too are made in the lymph glands and nodes, but they are phagocytic.

Platelets or **thrombocytes** are small fragments of cells and are formed in the bone marrow by the disintegration of certain cells. There are about 400,000 per mm^3 of blood; they have no nucleus and are important in blood clotting.

Functions of blood or circulatory system

It is often difficult to separate the functions of the blood from those of the circulatory system as a whole. The blood acts as a medium whereby materials can be exchanged with the tissue fluid surrounding the cells. The circulatory system is responsible for transporting substances to and from the cells making the exchange of materials possible. In addition the blood is very important in the prevention of infection.

The blood transports respiratory gases. Oxygen is transported from the lungs to the tissues as oxyhaemoglobin in the red cells. As a result of respiration in the tissues there will be a low concentration of oxygen and this causes the unstable oxyhaemoglobin to break down and release the oxygen which diffuses into the tissues. Carbon dioxide is transported to the lungs for excretion: it is produced from respiring cells and it diffuses from the cells and into the blood and it is carried to the lungs mainly as bicarbonate ions in the plasma. In the alveoli the carbon dioxide is released and diffuses from the blood into the alveoli.

Soluble digested food is transported from the ileum to the tissues; products such as glucose and amino acids are taken from the ileum to the liver in the hepatic portal vein; they may be used in the liver, stored, or released into the general circulation and taken to the tissues. Excess or unwanted amino acids are taken back to the liver for deamination.

Excretory products are transported to the excretory organs. Water may go to the kidney, skin or lungs. The main nitrogenous waste product is urea which is transported from the liver to the kidneys for excretion.

Hormones are transported from the endocrine organs in which they are produced to the target organs which they affect. For example insulin produced in the pancreas is released into the blood and carried to the liver where it controls the conversion of glucose to glycogen.

Heat is also distributed round the body. Chemical reactions release heat energy; the greatest number occur in the liver and muscles and the blood circulates the heat to less active parts of the body and excess heat is lost mainly through the skin.

Disease: prevention of infection

The body of man is continually being invaded by micro-organisms, particularly bacteria and viruses. Most bacteria are harmless but some cause diseases such as diphtheria, tonsillitis, tetanus, tuberculosis and typhoid. Viruses cause diseases such as poliomyelitis, influenza, the common cold and measles. These organisms enter the body mainly through the nose during inhalation, through the mouth when feeding and drinking, and through wounds in the skin. Symptoms of a disease can be caused by the foreign protein, for example of the bacteria itself, by products produced by the infected tissue, or by poisonous chemicals called **toxins** which are excretory products produced by bacteria.

The skin provides an effective barrier against the entry of micro-organisms, but when the skin is broken blood clotting is not only essential to prevent excessive loss of blood from the body, it is also important in minimising the entry of micro-organisms.

Blood clotting is a very complex process but the basis is the conversion of the soluble blood protein **fibrinogen** into an insoluble fibrous molecule called **fibrin** which forms a mesh to trap the blood cells and plasma. In the blood plasma is an inactive protein called **prothrombin**; it is made in the liver and vitamin K is essential for its production. When the skin is broken the platelets, on exposure to the air, and the damaged tissue, produce the enzyme **thrombokinase** (thromboplastin) which converts prothrombin to thrombin. **Thrombin** is usually regarded as an enzyme, and in the presence of calcium salts converts fibrinogen into fibrin. Fibrin is in the form of long fine threads in which the blood cells and plasma become entangled and a blood clot is formed. The clot slowly shrinks and a clear yellow fluid is released from it; this is **serum**, the name given to plasma when the fibrinogen has been removed from it. During the normal circulation of blood several substances prevent it from clotting; one is heparin produced by the liver.

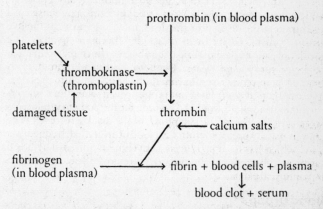

Phagocytis is very important in preventing infection. It is carried out mainly by polymorphs which can pass through capillary walls into the tissue spaces. They can move to sites of infection and engulf micro-organisms and in this way the infection is localised; bacteria which do enter the general circulation are acted upon by stationary white cells in the lymph glands and nodes. An accumulation of leucocytes and bacteria forms **pus**.

Antibodies are proteins which are produced by cells derived from lymphocytes, in response to the entry into the body of a foreign protein, called an **antigen**. Antibodies can act in several ways: **opsonins** adhere to the surface of the micro-organism making them more susceptible to phagocytosis; **agglutinins** cause them to clump together so they cannot invade tissues; **lysins** cause them to disintegrate. **Antitoxins** are a type of antibody; they combine with the poisonous toxins to produce harmless substances.

Immunity

The production of antibodies in response to the presence of an antigen is the basis of immunity. After recovery from an infectious disease the antibodies may remain in the blood for a considerable time, and a second attack is unlikely as the antibodies are already present to destroy any new invasion of that micro-organism. It is also possible to take in small doses of micro-organisms and the effects of the disease may not be felt, or only very mildly so, but the production of antibodies is stimulated. This type of immunity is called **naturally acquired** immunity.

Artificially acquired immunity is a very important part of preventive medicine. **Active acquired** immunity is brought about by the inoculation into the body of a **vaccine**; this may consist of a mild strain of the disease, or the dead micro-organism, or the live micro-organism, or its toxin which has been inactivated. The inoculation causes the build-up of antibodies which may remain in the body for several years. **Passive acquired** immunity is obtained by injecting a serum containing antibodies into the body. This immunity is short lived but is often given when there is a risk of infection and there would be no time for the body to produce its own antibodies before the symptoms of the disease were felt. The serum is usually prepared from the blood of an animal, frequently a horse, which has been previously actively immunised. The blood plasma is collected and the fibrinogen removed leaving a serum containing antibodies.

Blood groups

The human population can be divided into four blood groups based on the reaction which occurs when the blood of different humans is mixed together. The groups are A, B, AB and O and in Great Britain they are present in 42 per cent, 8 per cent, 4 per cent and 46 per cent of the population respectively. The letters stand for the type of **agglutinogen**, an antigen, present on the surface of the red corpuscles; thus a person belonging to blood group A has agglutinogen A on their red cells. Antagonistic substances called **agglutinins**, which are antibodies, may occur in the blood plasma. The agglutinins are known as $\propto$ and β (or anti-A and anti-B). If blood containing an agglutinogen and the corresponding agglutinin are mixed, the red cells clump together or **agglutinate**, for example if blood is mixed containing agglutinogen B and agglutinin β the blood will agglutinate. This caused the death of many patients in early transfusions before the four blood groups were discovered.

Group	Agglutinogens on red cells	Agglutinins in plasma
A	A	β
B	B	$\propto$
AB	A and B	none
O	O	$\propto$ and β

Table 2. Distribution of agglutinogens and agglutinins

Persons belonging to Group O are known as **universal donors** since neither agglutinogens A or B are present so they can donate blood to any other group. Persons belonging to Group AB are known as **universal recipients** as they can receive the blood of any other group.

Rhesus factor In 1940, after investigations using the Rhesus monkey, it was discovered that 85 per cent of white people have an extra agglutinogen on their red cells known as the Rhesus factor; persons possessing it are described as Rhesus positive (Rh+) and the 15 per cent lacking it, Rhesus negative (Rh−). Transfusion from Rh− to Rh+ is safe at all times. However, if an Rh− recipient receives a transfusion of Rh+ blood, an antibody, the anti-Rh factor, is built up: there is no immediate danger but if a second Rh+ transfusion is given the blood will agglutinate. In addition, if an Rh− mother conceives an Rh+ child some of the Rh factor passes from the baby through the placenta into the mother's blood where the anti-Rh factor is produced. The first baby is not affected but if a second Rh+ baby is conceived the anti-Rh factor passes from the mother's blood, across the placenta, and into the embryo and the baby will be born with jaundice or be stillborn. Today this situation can be detected and appropriate actions taken to overcome it.

The circulatory system

The circulatory system consists of the heart, blood vessels and lymphatic vessels. The muscular contractions of the heart maintain a one-way circulation through the three types of blood vessel, arteries, capillaries and veins. The lymphatic vessels contain lymph and return it from the tissues to the blood.

The heart

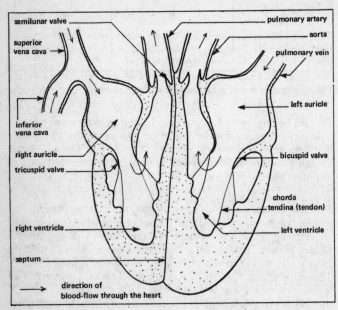

Figure 57. Vertical section through a mammalian heart

The heart is a hollow, muscular, pumping organ which is situated in the thorax, between the lungs and behind the sternum, and lying slightly to the left of the body. It is cone-shaped and is often said to be 'the size of a man's closed fist'. It is enclosed in a tough membrane, the **pericardium**, and receives oxygenated blood from the **coronary artery** which branches off the aorta. The heart is divided into right and left halves by a muscular septum and the two halves do not communicate. It consists of four chambers; on each side there is an upper thin-walled **atrium** or **auricle** which receives blood from the veins, and this leads into a lower, thicker walled **ventricle** which distributes blood to the arteries. De-oxygenated blood passes from the venae cavae into the right auricle, then into the right ventricle and through the pulmonary artery to the lung. Oxygenated blood enters the left auricle from

106

the pulmonary vein; it passes into the left ventricle and then into the dorsal aorta for distribution round the body. The wall of the left ventricle is considerably thicker than that of the right ventricle; this is because the left ventricle has to pump the blood right round the body whereas the right pumps the blood a much shorter distance to the lungs.

The heart possesses valves to prevent the backflow of blood. The **tricuspid valve** separates the left auricle and ventricle. The valves are held in position by **chordae tendinae** which are inelastic and pass from the valve to the ventricle wall. They prevent the valves being forced into the auricles when the ventricles contract. At the entrance to the arteries leaving the heart are **semi-lunar** or pocket valves.

Cardiac cycle The function of the heart is to maintain a constant circulation of blood through the body and its action consists of a series of events known as the cardiac cycle. Each cycle or heart beat lasts approximately 0·8 seconds; therefore in a 'normal adult' it occurs about 74 times per minute.

In each cardiac cycle the heart undergoes contraction (**systole**) and relaxation (**diastole**) and each lasts 0·4 seconds. Considering the right side of the heart, the sequence of events is as follows: during diastole (0·4 seconds) blood from the venae cavae enters the right auricle. Pressure on the tricuspid valve opens it and the blood flows into and fills the right ventricle. When the auricle and ventricle are full of blood the right auricle contracts and expels its blood into the right ventricle: this is **auricular** (atrial) **systole** and it lasts for 0·1 seconds. This is followed immediately by **ventricular systole** lasting 0·3 seconds, in which the right ventricle contracts and forces the blood through the semi-lunar valves and into the pulmonary artery. The tricuspid valves are closed at this point to prevent blood flowing back into the auricles. Systole and diastole occurs simultaneously in the right and left sides of the heart.

The cardiac cycle is started in the **pacemaker** in the right auricle, and the heart will continue beating rhythmically even after its nerve supply has been cut. However, the rate at which the heart beats is controlled by nerves and it receives two nerves through the autonomic nervous system. These are the sympathetic and vagus nerves and when stimulated they cause opposite effects: stimulation of the vagus nerve slows down the heart beat rate, while stimulation of the sympathetic nerve increases it. Heartbeat rate increases during exercise to enable more oxygenated blood to reach the tissues so the respiratory rate can be increased and more energy is released. Release of the hormone adrenaline also occurs in this situation (see page 147).

Arteries, capillaries and veins

Arteries always carry blood away from the heart. The artery wall is composed of three layers, an outer protective fibrous layer, a thick muscular middle layer which contains elastic tissue, and a very thin lining layer, endothelium. The thick muscle and elastic layer enables the heart to withstand the surges of high pressure caused by the heart beat. The largest artery, the aorta, has a considerable amount of elastic tissue but as arteries get further from the heart the amount diminishes. Arteries do not contain valves; in veins these prevent the backflow of blood but this is unnecessary in arteries as a rapid forward flow of blood is maintained by the contraction of the heart and by the squeezing effect of the recoiling elastic tissue. Most arteries lie deep in the body where they are protected; death can occur rapidly if an artery is severed as the blood spurts out each time the left ventricle of the heart contracts. With the exception of the pulmonary artery which passes from the heart to the lung, arteries carry oxygenated blood with very little carbon dioxide. Arteries possess a pulse. The **pulse** is a wave of distension felt in an artery

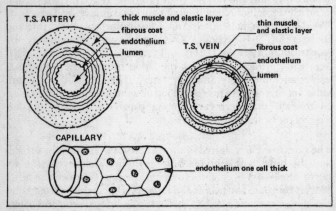

Figure 58. Structure of an artery, vein and capillary

108

wall owing to the contraction of the left ventricle forcing blood into an already full aorta; it can be felt where an artery comes near to the surface, for example in the radial artery in the wrist, and is felt by pressing the artery against a bone. Blood pressure in arteries is high but as they near the tissues they branch repeatedly, the blood pressure is reduced and the flow rate slowed down.

Capillaries are microscopic vessels which run between all the cells of the body. They receive blood from small arteries, called arterioles, and distribute it to small veins, called venules. The wall of the capillaries is only one cell thick and is made of pavement epithelium; this enables substances to be exchanged between the blood and tissue fluid.

Veins carry blood towards the heart. They are composed of the same three layers as the arteries but the muscle layer is very much reduced and there is very little elastic tissue as they do not have to withstand any surges of high pressure from the heart. As veins are

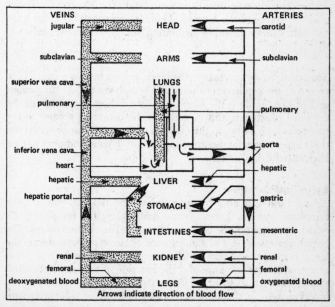

Figure 59. Main blood vessels of a mammal

109

away from the direct influence of the heart beat the venous pressure is low and **valves** are present along the veins to prevent the backflow of blood. Also in the limbs the blood has to be returned to the heart against gravity and its forward movement is aided by the contraction of the skeletal muscle around the veins. Veins lie nearer the surface of the body than arteries and the blood flow along them is slow and smooth. With the exception of the pulmonary vein which passes from the lungs to the heart, veins carry deoxygenated blood which contains much carbon dioxide.

Exchange of materials between capillaries, tissue fluid and lymphatics

When blood reaches the arterial end of a capillary its pressure is high because of the pumping action of the heart, and the fluid part of the blood is forced through the thin capillary walls and becomes part of the tissue fluid which bathes the cells. **Tissue fluid** is very similar to blood plasma but has fewer proteins, and it contains glucose, amino acids, salts and oxygen; the cells extract the substances they require for their own metabolism and excrete urea and carbon dioxide into the tissue fluid. The narrow capillaries offer resistance to the flow of blood, and its movement along the capillaries slows down, so facilitating diffusion from the plasma to the tissue fluid.

The tissue fluid can return to the blood system by two routes; it can return to the capillaries or pass into lymph vessels. At the venous end of the capillary the blood pressure is low as a result of the capillary resistance, and in addition the osmotic pressure of the plasma is high owing to the removal of the other constituents, and this causes tissue fluid containing carbon dioxide and urea to pass back into the capillaries.

Lymphatic system

Tissue fluid can also be returned to the bloodstream via the lymphatic system. Excess tissue fluid enters the lymph vessels which run between the cells, and becomes part of the **lymph**. The lymph vessels unite and run into one of two lymphatic ducts: the larger is the thoracic duct which receives lymph from the lower half of the body, the left arm, and the left side of the head, neck and thorax; the lacteals from the ileum also enter it. The ducts join the blood system in the neck at the junction of the subclavian and jugular veins.

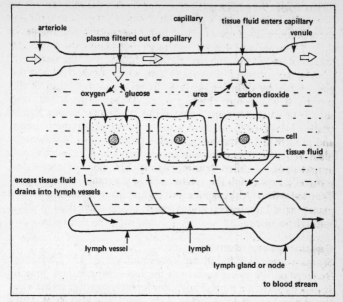

Figure 60. Relationship between the blood plasma, the tissue fluid and the lymph

Lymph vessels are very similar in structure to veins. The lymph flows in one direction and its forward movement is dependent on valves to prevent its back flow, on muscular activity and by pressure of the lymph. Lymph is a fluid very similar in composition to blood plasma, but it contains fewer proteins and more lymphocytes which are produced in lymph glands and nodes.

Lymph nodes are found in groups along the lymph vessels; they are in various parts of the body but are abundant in the neck, groin and in the armpit. In addition there are **lymph glands** such as the tonsils and spleen. The nodes and glands are important in preventing infection: they contain white cells which 'trap' micro-organisms in the lymph passing through them; they produce lymphocytes which enter the lymph and thus the bloodstream; they also produce antibodies.

Transport in plants

Plants require a supply of water and mineral salts, and the gases carbon dioxide and oxygen. In flowering plants the gases are mainly absorbed through the stomata of the leaves, the water and mineral salts through the root system. Once inside the plant the gases move by diffusion, always dissolving in the water of the moist cell wall before entering a cell. Water, mineral salts and soluble products are transported in special vascular tissues. Xylem tissue has evolved to transport water and mineral salts up the plant, and phloem tissue transports soluble products up and down the plant. The movement of soluble products, such as sucrose and amino-acids, is called **translocation**. The flow of water and mineral salts through the xylem is called the transpiration stream. **Transpiration** is the evaporation of water from the aerial parts of a plant (mainly the leaves) and it results in a massive flow of water through the plant.

Gas uptake and stomata

Stomata are pores found on leaves and stems. They are most frequent on the lower surface of the leaves where there may be several hundred per mm^2. They allow for exchange of gases between the atmosphere and the inside of the plant, and also allow water vapour to escape from the plant. As long as the stomata are open the plant will have an adequate supply of carbon dioxide for photosynthesis, but also runs the risk of wilting, or possibly dying, owing to a rapid loss of water which is not replaced. This problem is in part coped with by the controlled opening and closing of the stomata. Stomata tend to be open in the light and closed in the dark. This means that when photosynthesis cannot occur, and carbon dioxide is not required, water loss is prevented by the stomata being closed. There are occasions, for example during severe water shortage, when the stomata will close even in the light. Several theories have been proposed to explain the mechanism of stomata movement but none of them is completely satisfactory. The size of the stoma depends on the shape of the guard cells of the epidermis, which lie on either side of it. Guard cells are unusual, in that unlike other epidermal cells, they generally contain chloroplasts. They also have irregularly thickened cell walls. The form of the thickening of the guard cell walls means that when the turgor pressure inside them increases, by water being drawn in from neighbouring cells, the stoma opens. This is because the thicker cell wall, lining the stoma, is less elastic. When water is withdrawn from the guard cells the stomata will close.

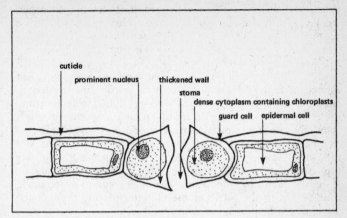

Figure 61. T.S. of stoma and guard cells

Transpiration

This results in vast quantities of water being lost from the aerial parts of land plants. For example, an acre of beech wood may lose up to $16,500 \text{ dm}^3$ of water on a warm day. Only a tiny fraction of the total volume of water passing through a plant is used for photosynthesis. This loss of water is an evil which the plant has to tolerate, the unfortunate fact being that a leaf structure which allows photosynthesis to take place efficiently also allows rapid water loss to occur. It is therefore debatable whether transpiration serves any useful function, but ones which have been suggested are that it cools the leaf in hot conditions, and that by producing the transpiration stream provides a method by which mineral salts are transported within the plant. Transpiration may be demonstrated by using cobalt-chloride paper. When dry the paper is blue, but it turns pink on becoming damp. A piece of blue paper is firmly attached to the surface of a leaf. After a short while the paper turns pink, indicating the release of water. This method may also be used to show which surface of the leaf (the upper or lower) loses most water, by recording the time it takes for the paper on each surface to become pink.

The external factors which affect transpiration are: 1. **Temperature** The higher the temperature the more rapidly water will evaporate from the moist cell walls of the leaf. 2. **Air movements**

113

When air is moving rapidly over the surface of the leaf, evaporation will also proceed rapidly. This is because water vapour will not be accumulating around the leaf and there will be no tendency for the air to become saturated. 3. **Humidity** When air is already carrying a large amount of water vapour (high humidity), evaporation will be slow as the air will quickly become saturated around the leaf surface. 4. **Light** This acts indirectly by causing the stomata to open and so allowing evaporation to occur. In order to demonstrate the effect of these factors on transpiration an experiment which shows the rate of transpiration must be used. A **potometer** is used to measure the rate at which water is taken up by a plant, and this water uptake is assumed to be equal to water loss. In fact this is slightly inaccurate since some of the water may be stored or used by the plant, but this volume is so small, compared with the total volume of water passing through the plant, that it may be ignored.

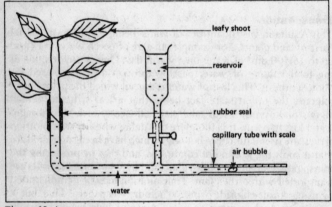

Figure 62. A potometer

The cut end of a leafy shoot is placed in the airtight rubber seal. This is done under water to prevent any air bubbles forming in the apparatus. The apparatus is left in the conditions being investigated for some time, and if functioning correctly an air bubble will enter the open end of the capillary tube. This is a result of water being drawn through the apparatus to replace that being lost by the plant. The position of the air bubble is adjusted to the beginning of the scale by opening the reservoir, and the rate of transpiration may then be measured by timing the movement of the air bubble along the scale. The experiment may be repeated under different conditions so that the rates of transpiration may be compared.

Movement of water across the leaf

As a result of transpiration, cells in the leaf which adjoin air spaces will tend to lose water from their vacuoles. Consequently the osmotic pressure of these cells will rise since the sap has become a stronger solution due to water loss. In addition the wall pressure will drop because of the decrease in volume of the cell contents. These two changes will produce an increase in the suction pressure of the cell. These changes are represented by the formula $SP = OP - WP$ (see page 37). Water will then move by osmosis from the neighbouring cells with a lower suction pressure. This produces a chain reaction across the leaf, water eventually being withdrawn from a xylem vessel. As a result of transpiration there is a continuous flow of water through the leaf, from the veins, through the cells to the moist cell walls which are losing water by evaporation.

Movement of water through the xylem

The removal of water from the veins of the leaf produces a tension or pull on the remaining water in the xylem vessels. This pull is sufficient to pull water up trees 100 metres high, and the column of water remains continuous, without any breaks throughout its length. It has been suggested that the water column does not break up, despite the enormous tension it is under, because water molecules in very thin tubes are so strongly attracted to each other. This is known as the cohesion theory. The transpiration stream carries water from the xylem vessels of the root, up the stem into the leaves. It may be demonstrated that xylem vessels transport water by placing the cut end of a leafy shoot in water stained with a dye such as methylene blue or eosin. After a short time the stem is sectioned and it is found that only the xylem vessels have become stained.

Movement of water through the root

Root systems serve to carry out two main functions, first that of anchoring the plant in the ground, and secondly that of absorbing water and mineral salts from the soil. As a result of cell division and, elongation just behind the root tip, the tip of the root is pushed downwards and outwards through the soil, forming a spreading system. This means that fresh areas of soil are constantly being used as a water supply. Water exists in drained soil as a thin film around the soil particles, and most of this water is absorbed by the **root hairs** of the root. These are finger-like projections produced by the superficial cells just behind the region of elongation.

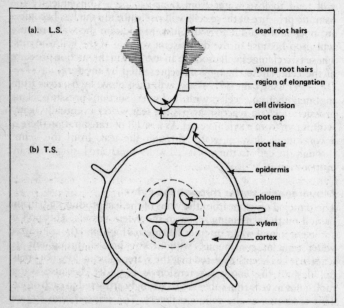

Figure 63. (a) L.S., and (b) T.S. of root hair region

The root hairs produce an enormous surface area through which a plant may absorb water and mineral salts. The fineness of the root hairs means that they can penetrate between the soil particles and come into contact with the film of water. The root hair cells do not secrete any waterproof substance. The soil water is a very weak solution and the root hairs contain the cell sap, which is a strong solution. Water will enter the root hair cells by osmosis, but the semi-permeable plasma membranes will prevent large molecules, such as sugar and amino-acids, from leaking out of the cell. As water enters the root hair cells it will produce a drop in their suction pressure by producing a decrease in osmotic pressure and an increase in the wall pressure. Again, this may be represented by $SP = OP - WP$. This will result in water then moving into neighbouring cells by osmosis and so on across the cortex of the root. The water can then enter the xylem vessels in the centre of the root, and be drawn up to the leaves in the transpiration stream.

The account just given of water movement through the root is certainly a simplification of the situation. It is possible that water may be actively drawn into roots by some method other than osmosis, which requires energy. Once inside the root, water may not only move just by osmosis from the vacuole of one cell to another, but possibly also move through the cytoplasm or through the cell walls of the root cells. There is also some evidence to suggest that the roots may pump water actively into the xylem. This evidence is the phenomenon known as root pressure. This occurs most frequently in spring when the cut ends of some stems exude water for a considerable time, indicating that water is under pressure in the xylem. The significance of this process is still not certain.

Osmo-regulation in plants

Plants which live in conditions where water is unavailable, either for long or short periods, show a number of adaptations which are thought to help overcome this problem. In temperate climates water tends to be limited only during the winter, when the ground is frozen or very cold. Annuals survive over winter only as seeds, which are protected by their low water content and hard testa. Many trees are **deciduous**, losing their leaves in autumn (e.g. ash and oak). This means that they reduce enormously the surface area from which water may be lost. Since the light intensity and temperature are low, and photosynthesis would be very slow, there would be little advantage in maintaining their leaves. The formation of buds, which are condensed protected shoots, at the end of the growing season, means that leaves may be rapidly produced as soon as conditions are suitable the following spring. Some trees, such as holly and pine, retain their leaves over winter and are described as **evergreen**. These plants tend to have a number of leaf adaptations, such as a very thick waxy cuticle, or a small number of stomata often sunken into pits or grooves, which should reduce the rate of transpiration.

In regions with a more severe water shortage plants may produce very deep root systems to penetrate low-lying reserves of water, or alternatively produce an extensive superficial root system, which will absorb water however brief the period of rainfall. It is in desert regions that the most extreme and perhaps familiar adaptations are to be found. **Succulent** plants store water, frequently in swollen stems, and reduce their leaves to spines. The plant is covered by a thick waxy cuticle to reduce transpiration. The cacti, such as the prickly pear, illustrate these features well. In general, plants are much more tolerant of changes in their water content than animals.

Movement of mineral salts

Plants absorb a variety of mineral elements through their root hairs, including nitrogen, phosphorus, sulphur and potassium. These ions are not absorbed passively with the water but enter the plant by some other method. These elements enter the plant in the form of separate ions. For example calcium phosphate enters in the form of calcium, (Ca^{++}) and phosphate (PO_4^{3-}) ions. Inside the plant the elements are often in higher concentration than outside, and some elements are more concentrated than others. This indicates that a selective process is occurring which involves the active uptake of salts using energy. It is not known how the salts move across the root, but once in the xylem vessels they are carried up the plant, along with the water, in the transpiration stream. The pathway of the salts up the plant was first established by ringing experiments. When a ring of bark and phloem is removed from the stem of a woody plant, leaving only the xylem tissue, it is found that the movement of salts is unaffected. Recently the use of radioactive isotopes has produced more exact information concerning the movement of mineral salts. It is possible to separate the xylem and phloem with waxed paper without damaging the plant. This prevents the sideways movement of substances between the two tissues. When a radioactive element, such as phosphorus, is fed to the roots of a plant treated in this way, the phosphorus is found to be restricted to the xylem. When the xylem and phloem are left intact the radioactive element is found in both tissues, indicating that normally mineral salts move out of the xylem sideways into the phloem. The mineral salts are mainly used at the growing points of a plant, and in photosynthesising cells for a variety of functions. Plants require the nitrates, phosphates and sulphates of potassium, magnesium, iron and calcium, and also minute traces of manganese, zinc, boron and cobalt. Nitrogen, phosphorus and sulphur are needed for the synthesis of molecules such as proteins, nucleic acids and A.T.P.; magnesium is needed as it forms part of the chlorophyll molecule; elements such as potassium and calcium are important in maintaining the correct environment for the metabolism of the cell to proceed efficiently. Some of the trace elements are important in allowing specific enzymes to function.

Translocation

This involves the movement of the products of photosynthesis. Translocation may take place upwards or downwards since substances such as amino-acids and sugars must be transported to all cells which are unable to photosynthesise. Regions which have the greatest demands for these substances are sites of growth, such as apices, and storage organs, such as bulbs and tubers. Translocation occurs within the phloem tubes. Ringing experiments, in which the bark and phloem are removed from a tree, result in sugars accumulating above the ring. Again, more specific information has been obtained about translocation by the use of radioactive isotopes. When plants are fed carbon dioxide containing radioactive carbon (^{14}C) this becomes built up into the products of photosynthesis. This radioactivity is only detected in the phloem of the stem when sections of the stem are placed on a photographic film. The mechanism of translocation is not yet understood. Any theory must explain the relatively rapid speed of translocation, and how substances can move up and down in the phloem simultaneously. It is believed that some type of active transport is involved, since anything which interferes with energy release in the phloem interferes with translocation.

Key terms

Artery Carries blood away from the heart; thick, muscular, elastic walls.

Capillary Allows exchange of materials between the blood and tissues; thin walled.

Lymph Similar to plasma but fewer proteins and more lymphocytes.

Plasma Pale yellow alkaline fluid in which corpuscles are suspended and substances dissolved.

Pulse Wave of distension felt in an artery wall.

Stoma Opening in the epidermis of a leaf or stem; size regulated by guard cells.

Tissue fluid Bathes the cells; derived from the plasma.

Translocation Movement of soluble products in the phloem.

Transpiration Evaporation of water from the aerial parts of plants.

Vein Carries blood towards the heart; thin walls with little elastic. Valves present along length.

Chapter 5
Excretion, Osmo-Regulation and Temperature Regulation

Excretion is the removal from the body of the waste products of metabolism. Metabolism is the name given to all the complex chemical reactions which occur continuously in living cells. These processes produce waste products (excretions) which are poisonous in varying amounts, and if they were allowed to accumulate, the body would cease working efficiently. It is often impossible to tell whether certain substances, for example water, have been produced in metabolism or whether they were absorbed from the alimentary canal and were not required by the body. In addition, therefore, excretion has come to include the removal from the body of excess substances that have passed across cell membranes and have formed part of the protoplasm.

Excretion should not be confused with egestion or secretion. Egestion is the removal from the body of the undigested products of digestion; these substances have never crossed a living membrane to be part of the cellular constituents. Secretion is the production of useful chemical substances by living cells. Sweat produced by mammals can be regarded as a secretion or excretion. The water component has an important function in temperature regulation, whereas the sodium chloride and small amount of urea dissolved in it are excretory products.

Osmo-regulation is the process by which the osmotic pressure of the blood and body fluids is kept constant. This involves keeping the water and salt content within narrow limits and this is achieved by controlling how much water and salt is allowed to enter and leave the cells and body fluids.

Excretion and osmo-regulation are very closely linked, and in the mammal the kidney performs both functions.

Excretory products

The main ones in animals are carbon dioxide and water from respiration, nitrogenous compounds, salts, and bile pigments from the breakdown of haemoglobin in 'worn out' red corpuscles. Plants do not excrete waste nitrogenous substances but do produce carbon dioxide and water from respiration, and oxygen from photosynthesis. The type of nitrogenous waste excreted differs in different organisms. Amino-acids which are in excess after absorption, or which have been released by the breakdown of protoplasm in damaged or worn-out cells, undergo **deamination**. This process, which occurs in the liver of mammals, involves the removal of the amine group from the amino-acid and its formation into ammonia. **Ammonia** is highly toxic and whether it is excreted as such or converted to **urea** or **uric acid** depends on the environment of the organism and how much water is available to it.

In many aquatic organisms such as *Amoeba* and many amphibian larvae (tadpoles) the ammonia is quickly diffused out all over the body surface into a large volume of water where it is soon dispersed and its concentration negligible. Urea is the main nitrogenous waste product in terrestrial and semi-terrestrial organisms such as mammals and frogs where there is a moderate amount of water in the environment. Ammonia would be highly toxic so it is combined with carbon dioxide in the liver and converted to urea. Urea is soluble, but less toxic than ammonia allowing it to be concentrated and stored in the body and released at intervals. Uric acid is the main nitrogenous waste product in terrestrial organisms and is especially suitable for those living in conditions where there is little water available. It is insoluble and non-toxic and in insects is excreted as dry pellets, and in reptiles and birds as a thick paste.

Excretory organs

In man the **lungs** excrete carbon dioxide and water from respiration (see page 94) and the **kidneys** excrete water, salts and nitrogenous waste products. In addition water, sodium chloride and a little urea and lactic acid are lost from the **skin** in sweat, but sweat production occurs in response to increase or decrease in body temperature and not as a response to changes in the chemical composition of the body fluids. The **liver** can also be said to have an excretory function when releasing bile pigments into the intestine.

The mammalian kidney

A mammal possesses two kidneys which are each enclosed in a transparent membrane and attached to the back of the abdominal cavity; they are red-brown oval structures with an indentation on their inner surface. Each kidney receives oxygenated blood from a

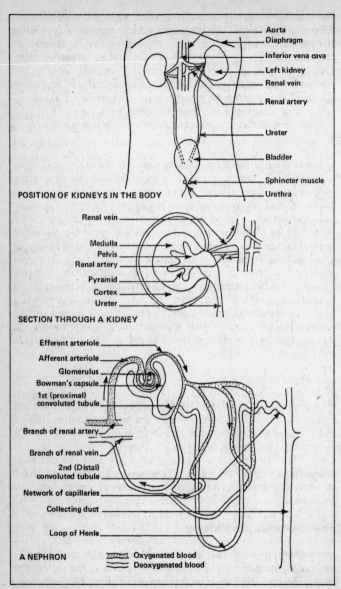

Figure 64. *The position of the kidneys in the body and their structure*

122

renal artery which branches off the aorta, and deoxygenated blood leaves by a renal vein which joins the inferior vena cava. A tube, the ureter, runs from each kidney to the base of the bladder, from which the urethra leads to the outside.

In vertical section the kidney is seen to consist of a dark outer region, the **cortex**, and a lighter inner region, the **medulla**. The inner border of the medulla is extended into several conical pyramids which project into a space, the pelvis, which is continuous with the ureter. In the cortex and medulla there are about one million filtering units called **nephrons**. At the beginning of each nephron and situated in the cortex is a cup-shaped **Bowman's capsule** which has very thin walls of squamous epithelium (see page 15). Within it is a knot of blood capillaries, a **glomerulus**; this receives blood from the wide afferent arteriole which branches from the renal artery, and it distributes blood to the much narrower efferent arteriole. From the Bowman's capsule runs the **first** or **proximal convoluted tubule** which leads into a U-shaped **loop of Henle**; this dips into the medulla and then returns to the cortex where it continues into the **second** or **distal convoluted tubule** which leads into a collecting duct or tubule. This duct opens at the tip of a pyramid and its contents pass into the pelvis and down the ureter to the bladder. The efferent arteriole, which carries oxygenated blood, branches into a network of capillaries over the tubules.

Functioning of the kidney

It is in the nephrons and their associated blood vessels that the composition of the blood is adjusted and urine is produced: this is achieved by two processes, pressure or ultra-filtration and selective reabsorption.

Ultra-filtration or **pressure filtration** is carried out in the Bowman's capsule. The afferent arteriole leading into the glomerulus is wider than the efferent arteriole leading out, and this sets up a high pressure within the glomerulus. This pressure causes small molecules in the blood, such as glucose, amino-acids, salts, water and urea, to filter through the capillary walls and into the cavity of the Bowman's capsule. Blood cells and protein molecules, such as fibrinogen, remain in the blood capillaries as they are too large to be filtered. This process is passive and does not require the expenditure of energy.

If all the substances passing into the Bowman's capsule (the glomerular filtrate) were to form urine, many valuable substances would be lost from the body. A process called **selective reabsorption** occurs in the tubules, where substances filtered into the Bowman's capsule are reabsorbed back into the blood capillaries which pass over them. In the first convoluted tubule glucose, amino-acids and water are reabsorbed, in the loop of Henle water is reabsorbed, and in the second convoluted tubule salts and water are reabsorbed. Reabsorption often occurs against a concentration gradient so it must involve active transport and the expenditure of energy; this is supported by the fact that reabsorption can be slowed down or stopped by the use of chemicals, such as cyanide, which are known to inhibit respiration.

The liquid entering the collecting duct is now urine and it passes down the duct where more water is reabsorbed. It is in the second convoluted tubule and collecting duct that the concentration of the blood is regulated. If the blood is too dilute, for example after drinking a large amount, less water is reabsorbed into the blood and the urine is dilute. If the blood is too concentrated, for example after profuse sweating, more water is reabsorbed and a concentrated urine is produced. The urine passes out of the collecting duct and into the pelvis and it is then passed down the ureters to the bladder where it is stored. At the base of the bladder is a sphincter muscle and this relaxes at intervals and the urine is released through the urethra to the outside. This process is known as **micturition**. Blood leaving the kidney in the renal vein will contain less oxygen and glucose but more carbon dioxide, as a result of energy expenditure during selective reabsorption, and less water, salts and nitrogenous waste as a result of excretion.

Urine consists of water, salts, waste nitrogenous substances such as urea and uric acid, and traces of other substances such as hormones, water-soluble vitamins and drugs. Glucose is completely reabsorbed in the tubules and none should be present in the urine. The glucose level in the blood is kept constant by the hormone insulin, which converts any excess glucose to glycogen (see page 74). If a person is suffering from the disease **diabetes melitus**, not enough insulin is produced; this causes the level of glucose in the blood to rise and when it reaches a certain level it is released in the urine.

Water balance and osmo-regulation

The body gains water by eating and drinking, and water is also a product of tissue respiration. Water leaves the body in urine, sweat, exhaled breath and faeces. These gains and losses will cause fluctuations in the composition of the blood and body fluids. Similarly the amount of salt taken in with food and the amount given out in sweat will also fluctuate. The osmotic pressure of the blood and body fluids should be maintained within narrow limits and this is achieved by controlling how much water and salt is reabsorbed into the blood from the second convoluted tubule and collecting duct. The kidney also excretes acid salts to keep the pH of the blood alkaline. The blood cannot influence the kidney tubules directly but does so via hormones.

An increase in salts in the blood, or loss of water owing to, for example, excessive sweating, will cause the osmotic pressure of the blood to rise. When this blood passes through the hypothalamus in the brain this is detected and the pituitary is stimulated to release a hormone called **antidiuretic hormone (ADH)** into the blood. In the kidney ADH causes an increase in the amount of water reabsorbed into the blood, so more water is retained in the body and a more concentrated urine is produced. A decrease in salts in the blood or an increase in water, due to drinking a large quantity of it, will cause the osmotic pressure of the blood to fall; this too is detected by the hypothalamus as the blood passes through it, and causes the pituitary gland to release less ADH. As a result less water is reabsorbed into the blood from the kidney tubules and a more dilute urine is produced.

Contractile vacuoles

These structures are present in many freshwater protozoans, for example in *Amoeba* and *Paramecium*, and they are thought to be solely osmo-regulatory organs. The cell membrane surrounding these organisms is semi-permeable, and as the osmotic pressure within the organism is greater than the osmotic pressure of the environment in which it lives, water continuously enters the cell by osmosis. This excess water is secreted into the contractile vacuole which expands and when it reaches a certain size discharges its contents to the outside and another contractile vacuole forms. There are numerous mitochondria surrounding the contractile vacuole, indicating that energy is required for its action. If salt is added to the fresh water this increases its osmotic pressure and the discharge rate of the vacuoles decreases.

Temperature regulation

Mammals and birds can maintain a constant body temperature which is independent of the temperature of the environment. Such organisms are said to be **homoiothermic**, which means 'having the same temperature', but they are popularly described as 'warm blooded'. The maintenance of a constant body temperature is an important aspect of **homeostasis**. In man the body temperature remains at approximately 36·9°C which is the optimum temperature for enzyme action; enzymes control all the chemical reactions in the body and by maintaining a constant body temperature they are allowed to work at their most efficient rate at all times.

In contrast all invertebrates, fish, amphibia and reptiles, have a variable body temperature which changes with fluctuations in the temperature of the environment. Such organisms are said to be **poikilothermic**, which means 'having a variable temperature'. They are popularly described as 'cold-blooded' but in fact their blood is often far from cold and some of them have a temperature well above that of their surroundings. In general, however, their temperature is dependent on environmental temperature and it will be the same as or only a few degrees above it. A decrease in environmental temperature and subsequently in body temperature will slow down all the enzyme-controlled chemical process in the body and this may reduce the organism to a state of inactivity.

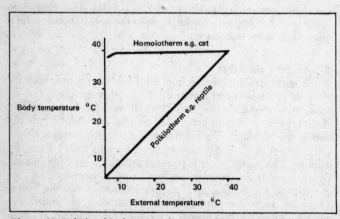

Figure 65. Relationship between the internal and external body temperature in a cat and a reptile

Skin and temperature regulation

The skin is an organ since it is composed of several types of tissue and is co-ordinated to perform specific functions, but it is unusual since it is very widespread and covers the whole surface. The skin consists of two layers, an upper epidermis, and a lower dermis. The structure and formation of the epidermis is dealt with on page 16. Sweat glands, sebaceous glands and hair follicles are situated in the dermis but are derived from the epidermis. The dermis is a layer of connective tissue containing fibres, blood capillaries, lymphatic vessels, sense organs and nerves. Beneath the dermis is a layer of subcutaneous fat, which acts as a food store and helps to reduce heat loss.

Apart from temperature regulation and food storage, the skin has many functions: it is impermeable and prevents desiccation, it protects the body against mechanical damage and the effects of the sun's radiation, and it prevents the entry of micro-organisms. It contains numerous sense organs so the organism is able to detect changes in its environment. It is an excretory organ, and it synthesises vitamin D. In addition, structures such as teeth, mammary glands, nails, whiskers and claws are modified skin structures. These functions will be dealt with in more detail in other parts of the book.

Heat gain and heat loss Chemical reactions which occur in all the cells of the body will produce heat. The greatest amount is produced in the liver and muscles and it is distributed around the body in the circulatory system. Heat is lost when any material is passed out of the body, for example during egestion, exhalation and urination, but the greatest heat loss is under the direct control of the brain and takes place from the skin when the body temperature is higher than the environmental temperature. Heat can be lost from the body by radiation, convection and conduction, and by the evaporation of sweat. To convert a liquid into a vapour, heat is required, and sweat evaporates using latent heat from the body and this reduces the body temperature.

Normally a balance is maintained whereby heat loss is equal to the heat gain. A man's temperature will vary in different parts of the body but a reading taken from under the tongue will always read about 36·9° C under normal conditions.

In attempting to maintain a constant body temperature two processes can occur for which the body must be able to compensate: these are overheating and overcooling.

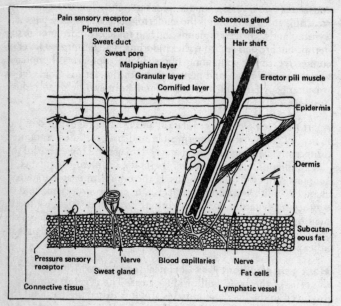

Figure 66. Section through the skin of man

Overheating

This can be caused by many external factors including vigorous activity, high environmental temperature, absorption of radiation from the sun, and disease. The hypothalamus in the brain is sensitive to the temperature of the blood flowing through it; if the blood temperature is a fraction above 'normal' this is detected and nerve impulses are sent to the skin and marked effects occur in an attempt to reduce the body temperature.

The surface arterioles and capillaries in the skin dilate (get wider), a process known as **vasodilation**, so more blood flows to the surface and heat is lost to the atmosphere by radiation and convection. Sweating is increased as secretory cells in the sweat gland absorb fluid from the blood capillaries and this 'sweat' passes up the sweat duct and on to the surface of the skin where it evaporates. Latent heat is taken from the body to evaporate the sweat so the body temperature is reduced. The evaporation rate is increased by air movements over the skin, but is decreased when the air becomes humid and the amount of water vapour it is holding increases. In addition the hairs are lowered by relaxation of the erector pili muscles so that they lie flat against the body. This means that very little air is trapped between the hairs; air is a poor conductor of heat, so heat insulation is reduced and heat is lost more readily by radiation and convection. The metabolic rate (rate at which chemical reactions are carried out) decreases so less heat is produced. This is why animals are generally less active in hot weather.

Overcooling

This is generally brought about by a low environmental temperature or inactivity. If the blood passing through the hypothalamus is a fraction below 'normal' this is detected and nerve impulses are sent to the skin and effects occur in an attempt to conserve heat.

The surface arterioles and capillaries constrict (get narrower) by a process known as **vasoconstriction**; as a result less blood flows to the surface so less heat loss occurs by radiation and convection; the blood is diverted to the deeper layers of the body. Sweat production is reduced, so keeping heat loss due to the evaporation of sweat down to a minimum. The erector pili muscles contract and this raises the hair;

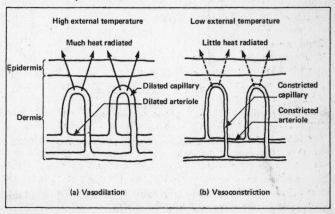

Figure 67. Vasodilation and vasoconstriction

129

as a result a thicker layer of air is trapped between the body and the atmospheric air and as air is a poor conductor of heat it helps to insulate the body and reduce heat loss. A similar response occurs in birds when they fluff out their feathers. In man the hair is so sparse that contraction of the erector pili muscles in an attempt to trap a thicker layer of air results only in 'goose pimples'. Shivering is a reflex action which occurs when the body temperature begins to drop; the muscles of the skeleton contract spasmodically and some heat is produced. Also the metabolic rate is increased, particularly in the liver, and this causes more heat to be produced.

Surface area and heat loss

As organisms get larger, then body surface increases by the square of its dimensions whereas the volume increases by the cube. Thus as organisms increase in size their surface area to volume ratio decreases and this means that small organisms have a larger surface area in relation to their volume (i.e. per unit of volume) than larger organisms. This can be demonstrated by considering three cubes of sides 1mm, 6mm and 10 mm.

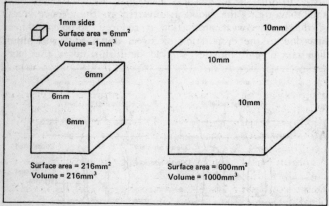

Figure 68. Surface area and volume of three cubes of differing sizes

In the 1 mm-sided cube there are 6 mm^2 of surface per mm^3. In the 6 mm-sided cube there is 1 mm^2 of surface per mm^3. In the 10 mm-sided cube there are 0·6 mm^2 of surface per mm^3.

Small birds and mammals lose heat very rapidly to their surroundings and they have large food demands and a high metabolic rate. In temperate climates, such as that of Great Britain, many small mammals such as fieldmice and hedgehogs hibernate through the winter to enable them to survive the period when heat loss would be high and food would be scarce. During hibernation the mammals sleep in a specially prepared nest. The body temperature drops considerably, often to a few degrees above that of the surroundings, the heart-beat rate and breathing rate is very low and all metabolic processes are reduced to a minimum. The organism uses its food store, predominantly fat, which was built up the previous autumn. At the end of hibernation the body temperature rises and the animal becomes active again.

Many organisms have evolved adaptations to enable them to reduce heat loss or heat gain, particularly in extreme climates. In cold climates there is a tendency for protruding parts of the body to be smaller to reduce heat loss; this can be seen in the Arctic fox. In contrast the Fennec fox which lives in the deserts of North Africa has very large ears to encourage heat loss.

Key terms

Deamination Removal of the amine group from amino-acids; in mammals it is converted to urea and excreted; the carbon residue is used in respiration or stored.

Egestion Removal from the body of undigested products of digestion.

Excretion The removal from the body of the waste products of metabolism.

Homoiothermic Maintaining a constant body temperature which is independent of the temperature of the environment.

Kidney Organ of nitrogenous excretion and osmo-regulation.

Metabolism All the chemical processes occurring continuously in living organisms.

Nephron One of millions of units in the kidney which filter the blood and in which urine is formed.

Osmo-regulation Process by which the osmotic pressure of the blood and body fluids is kept constant.

Poikilothermic Having a body temperature which is the same as or a little above that of the surroundings, and upon which it is dependent.

Secretion Production of useful chemical substances by cells.

Urine Solution of water, salts, waste nitrogenous substances which is produced in the kidney and stored in the bladder.

Chapter 6
Locomotion and Support

Some kind of movement can be recognised in most organisms. It may vary from movement of cytoplasm within a cell to movement of an organ, for example a leg or a leaf, or lastly to movement of the whole organism from one place to another; it is this change in position which is known as locomotion. Locomotion is much more a characteristic of animals than plants where it is confined to microscopic algae (such as *Chlamydomonas*) and motile gametes. This difference is related to their modes of nutrition: the green plant has autotrophic nutrition and carries out photosynthesis; the raw materials required for this process are present in the soil and surrounding air and therefore locomotion is unnecessary. In contrast animals have heterotrophic nutrition and require a ready-made organic food source, and they have to search for this, often over a wide area. Other advantages are gained by organisms which are freely mobile: they can avoid predators, seek out mates, and disperse to new areas.

With few exceptions locomotion is brought about by contraction of muscles acting against some kind of skeleton. However, in some of the lower organisms such as *Amoeba* and *Paramecium* specialised types of locomotion occur.

Three types of skeleton can be recognised in animals and they are important in support as well as locomotion. Earthworms and jellyfish possess a **hydrostatic skeleton** in which fluid in the body is under pressure and surrounded by muscles which contract against it. In **exoskeletons** the hard material is mainly confined to the outside of the organism and it encloses the soft body tissues. The arthropods, which include insects and crustaceans e.g. crabs, woodlice, possess an exoskeleton or cuticle, and in places the cuticle projects into the body to provide muscle attachment. The cuticle is inelastic and incapable of growth so that arthropods grow by a series of moults or ecdyses where the old cuticle is shed and replaced by a larger one which develops underneath (see page 166). Vertebrates possess an **endoskeleton**, where the hard material, bone or cartilage is within the body. These animals grow continuously until a certain size is reached and then they stop (this is discussed in more detail in chapter 8).

Skeleton in mammals

Mammals possess an endoskeleton composed mainly of bone. It can be divided into an **axial** skeleton, consisting of the skull, vertebral column, ribs and sternum, and the **appendicular** skeleton consisting of the pectoral (shoulder) girdle and the fore limb, and the pelvic girdle, and the hind limb.

The **skull** in man consists of twenty-two bones which apart from the lower jaw are fused together. It includes the cranium, which encloses the brain, and paired sense capsules. The auditory capsule encloses and protects the middle and inner ears, the nasal capsules enclose nasal cavities, and the orbits surround and protect the eyes. The upper jaw is fused to the floor of the skull, and it articulates with the lower jaw; both jaws bear teeth. At the posterior of the skull is a large opening, the foramen magnum, through which the spinal cord passes from the brain.

The **vertebral column** consists of thirty-three bones in man and forty-six in the rabbit. The individual bones are called **vertebrae** and they are specialised in different regions of the vertebral column for specific functions, but they typically have the same basic structure.

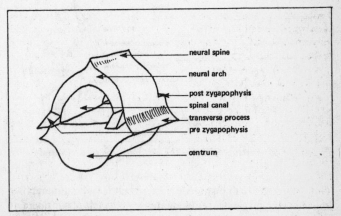

Figure 69. A typical vertebra

There are seven **cervical** vertebrae in man and rabbit and they support the neck; they are characterised by vertebrarterial canals which lie on either side of the neural canal and carry the vertebral arteries. The first two cervical vertebrae are specialised to carry the head; the first, the **atlas**, articulates with the skull and allows it to nod; the second, the **axis**, allows the head to rotate. For this purpose it possesses an odontoid peg which projects into the neural canal of the atlas. In man and rabbit there are twelve **thoracic** vertebrae which form an integral part of the rib cage. They possess facets between the adjacent centra and on the transverse processes for articulation of the ribs; they also have long neural spines for attachment of the shoulder and back muscles. In man the five **lumbar** vertebrae (seven in rabbit) are large and strong with well-developed zygapophyses and transverse processes for the attachment of the powerful back muscles. The five sacral vertebrae (four in rabbit) are fused together for extra strength to form the **sacrum** which is attached to the pelvic girdle; it has to transmit the full force applied by the hind limbs during movement to the rest of the body. There are four caudal vertebrae in man which are fused to form the **coccyx**; it is small and functionless and shows a progressive loss in processes; in the rabbit the sixteen caudal vertebrae support a tail.

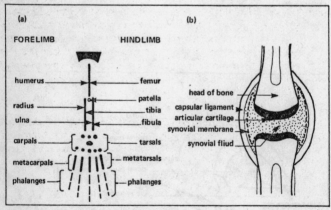

Figure 70. (a) Pentadactyl limb, (b) section through a synovial joint

The limb skeleton in mammals is based on the **pentadactyl** (five finger) plan typical of land vertebrates; it is capable of considerable adaptation according to whether it is a fore or hind limb or

134

according to the method of locomotion. In general the hind limb provides the thrust in forward movement and the fore limbs act as shock absorbers. The limbs are anchored in the body wall by the limb girdles. The **pectoral** or shoulder girdle consists of two bones on either side: the **scapula** or shoulder blade, a flat triangular bone which lies against the back of the rib cage, and the **clavicle**. The **pelvic** or hip girdle is formed from three bones fused together on either side; they meet in the mid line and are firmly attached to the sacrum. The whole structure is extremely strong and this is essential as it is subject to considerable force as it transmits the thrust of the hind limbs to the vertebral column.

Functions of skeletons

The functions of the skeleton can usually be grouped under the following headings: support and maintenance of body shape, protection, and locomotion.

Support
In aquatic organisms the water plays an important part in support and many invertebrates do not possess a skeleton. The part played by the skeleton in support in fish is not as important as in terrestrial vertebrates. On land support is achieved entirely by the skeleton; it raises the body off the ground and allows it to move across it. In addition many of the internal organs are suspended from the skeleton. In earthworms support and maintenance of body shape is achieved by the pressure of the fluids in the body cavity acting outwards against the muscular body wall.

Protection
The role of the skeleton in protection can be demonstrated most clearly in mammals; for example, the skull protects the brain and sense organs of the head, the vertebral column protects the spinal cord, and the rib cage protects the heart and lungs. In arthropods the cuticle protects the whole body.

Movement
Movement is usually achieved by the contraction of muscles acting across joints as levers with a low mechanical advantage. Bones are connected to each other by tough elastic **ligaments** and where two bones meet a joint is formed. The bones are moved relative to each other by muscles which are attached by inelastic tendons. **Muscles** work in **antagonistic pairs**, one

producing movement in one direction and the other producing movement in the opposite direction. For example, flexor muscles flex or bend the limb and extensor muscles extend or straighten it; similarly adductors move the limb towards the body and abductors away. Muscles can contract to about a half or two-thirds their resting length and this has the effect of altering the shape of the muscle (it becomes shorter and thicker) but not its volume. When muscles contract they pull on bones (they do not push). When it relaxes a muscle has to be pulled back to its resting length by another muscle contracting. At rest muscles are always slightly contracted and are said to be in a state of tension or tone and this is important in support.

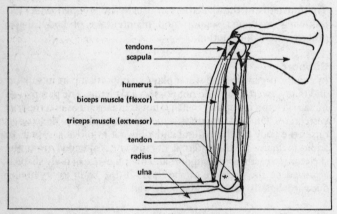

Figure 71. Antagonistic muscles of the upper arm

Muscle action can be demonstrated by considering an antagonistic pair of muscles such as the **biceps** muscle and **triceps** muscle of the upper arm. With reference to figure 71, if the palm is upwards, contraction of the biceps muscle raises the forearm; contraction of the triceps muscle and relaxation of the biceps muscle lowers the forearm. The biceps is the flexor muscle and the triceps is the extensor. One muscle of a pair is usually larger and stronger; in this example it is the biceps muscle as it has to lift the limb against the pull of gravity. Muscles act across joints as levers with a low mechanical advantage. The muscle contracts a short distance but as it is attached near the joint the movement at the end of the limb is greatly magnified. Efficient co-ordination of antagonistic muscles is essential and this is partly achieved by stretch receptors in the muscles and tendons (see page 149).

Joints can be classified according to their structure and the degree of movement possible. Immovable or **fixed** joints are found in the skull, pelvis and sacrum; the bones are bound to each other for extra strength and no movement is possible. Slightly movable joints occur between the vertebrae in the vertebral column.

The most common joints are **synovial joints** which allow considerable free movement. They are enclosed in a tough, fibrous capsular ligament which limits the amount of movement possible. The ends of the bones are covered with articular cartilage which is smooth and slippery and has a cushioning effect; this, together with the lubricating synovial fluid which fills the cavity of the joint, allows friction-free movement. The fluid is secreted by the synovial membrane which lines the inside of the joint. The bones may be held together by additional ligaments for extra strength.

Synovial joints are classified according to the degree and type of movement possible. **Ball and socket** or universal joints are found at the hip and shoulder and they allow movement in all planes, including rotation. The **hinge joint** found in the knee and elbow allows up to 180° movement in one plane only. **Gliding** joints are found where the flat surfaces of bones glide across each other; for example, between the bones in the wrist and ankle to give flexibility.

Locomotion in other vertebrates
Fish They have a smooth surface and streamlined shape to enable them to move easily through the water. The muscular tail is used to drive the fish forwards and this is achieved by alternate contraction and relaxation of muscles lying on opposite sides of the flexible vertebral column. The tail and body exerts a sideways and backwards thrust against the water and the resistance it offers pushes the fish sideways and forwards. The dorsal and ventral fins prevent the fish rolling, the paired pectoral and pelvic fins prevent pitching and act as brakes to slow down or stop the fish. Some fishes possess an air-filled swim bladder which allows them to remain at a certain depth in the water when they stop swimming. The amount of air in the bladder is altered so the fish can swim at different depths.

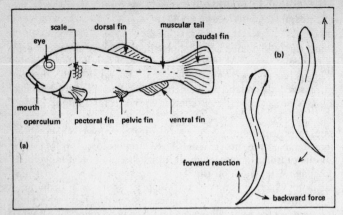

Figure 72. (a) A generalised fish, (b) swimming movements

Frogs The powerful, elongated muscular hindlimbs are adapted for swimming and leaping. When leaping the flexed limb is straightened, pushing the limb against the ground, and the thrust is transmitted through the pelvic girdle to the rest of the body; the short front legs are extended to absorb the shock of landing and then the hind limbs are drawn up to them. In water the webbed hind feet increase the surface area that can push against the water and the forelimbs are used to steer.

Birds They are well adapted for flight. They are streamlined and the forelimbs are modified into wings which are covered with feathers to increase their surface area. In addition birds are light due to their hollow bones and air sacs, and the large keel-shaped sternum or breast bone provides an anchorage for the strong pectoral wing muscles. The wing is an example of a pentadactyl limb but there is fusion of bones and a reduction in the number of digits.

Birds can carry out flapping flight, or can glide or soar. During flapping flight the pectoralis major muscle contracts and pulls the wings obliquely downwards and backwards giving the bird vertical lift and forward thrust. The feathers are held together during the downstroke to give maximum resistance against the air. To raise the wing the pectoralis minor muscle contracts and the feathers part to allow air to pass through them so that the wing offers little resistance to the air.

Movement in selected invertebrates

Amoeba *Amoeba proteus* is a single-celled protozoan which lives in fresh-water ponds. It has no fixed shape and moves by extending pseudopodia ('false feet') in the direction in which it is moving and withdrawing them at the opposite end. The cytoplasm is divisible into an outer ectoplasm and an inner endoplasm and the endoplasm exists in two states, as a thin, outer, firm layer of plasmagel, and an inner fluid region of plasmasol. This division of the endoplasm into plasmagel and plasmasol is questionable and the mechanism by which movement is brought about is still very uncertain. One possible explanation is that where a pseudopodium is to be formed the plasmagel changes to plasmasol and the change exerts a force which pulls the plasmasol towards this region and it pushes out the cell membrane and forms a pseudopodium. As the plasmasol streams into the new pseudopodium it fans out near the tip and gelates. At the 'temporary' posterior of the animal the plasmagel is changing to plasmasol and streaming forward.

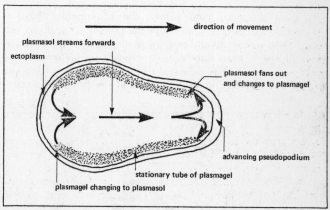

Figure 73. Diagram to explain one theory of movement in Amoeba

Paramecium caudatum This is a protozoan which inhabits fresh-water ponds. The whole of its surface is covered with minute hair-like structures called **cilia** and their co-ordinated beating causes movement. Each cilium is a fraction of a second ahead of the one in front in its beat, and each beat has two phases. In the effective stroke the cilium is stiff and pushes against the water causing forward movement, and this is followed by a recovery stroke when the cilium is limp so offering little resistance to the water and it moves the cilium to the position where the effective stroke can begin again.

Earthworm Movement is brought about by the contraction of circular and longitudinal muscles in the body wall which act on the fluid in the body cavity beneath. In addition there are four pairs of bristles or chaetae in every segment but the first and last, and they can be extended into the burrow to anchor the earthworm. To move, the circular muscles in the anterior contract, the segments become long and thin, and the anterior of the animal is pushed forward through the soil. The chaetae in this region are then pushed into the sides of the burrow, the longitudinal muscles contract, the circular muscles relax, and the segments behind the anterior end are pulled up. Several such alternate waves of contraction and relaxation of the muscles occur along the body.

Support in plants

The most advanced land plants, the angiosperms, have a well-developed shoot system above the ground. This allows the leaves to be spread out and exposed to the sunlight and available gases. It also permits flowers to be exposed for pollination and seeds for dispersal. The stem must have some means of support and this will vary depending on the size and life-span of the species.

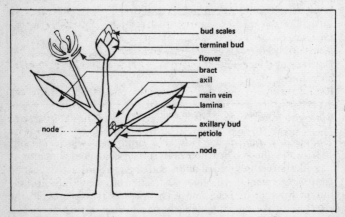

Figure 74. Shoot system in herbaceous plant

140

In herbaceous plants flexibility and strength are achieved by a combination of turgor pressure and supporting tissues. The cells of the pith when fully turgid push outwards and this force is restrained by the inelastic epidermis. The importance of this means of support in herbaceous stems is seen when plants which are deprived of water wilt. In the vascular bundles, xylem vessels and fibres, which contain lignin, add mechanical strength to the stem. In woody perennials support is entirely achieved by supporting tissues. Each year, as a result of cambial activity, a new layer of xylem is added to the stem. This results in the formation of large tree-trunks which consist mainly of 'wood' (xylem vessels and fibres).

Key terms

Appendicular skeleton Consists of the pectoral girdle and forelimb, and pelvic girdle and hind limb.
Axial skeleton Consists of the skull, vertebral column, ribs and sternum.
Ball and socket joint Allows up to 360° movement in all planes; found at the hip and shoulder.
Endoskeleton Hard material is on the inside of the body, e.g. skeleton of vertebrates.
Exoskeleton Hard material is on the outside of the body and encloses the soft body tissues; skeleton of insects.
Hinge joint Allows up to 180° of movement in one plane only; found at the elbow and knee.
Hydrostatic skeleton Fluid in the body is under pressure and surrounded by muscles which contract against it; found in worms.
Ligament Joins bone to bone.
Joint Structure formed where two bones meet.
Pentadactyl limb Five-fingered limb found in amphibia, reptiles, birds and mammals.
Pseudopodia 'False feet'; temporary structures which are extended during movement in Amoeba.
Synovial joint Freely movable; enclosed in a capsular ligament. The cavity contains synovial fluid secreted by a synovial membrane; heads of bones covered with articular cartilage.

Chapter 7
Co-ordination

Co-ordination in animals

The various physiological process which occur in living organisms are very closely linked and dependent upon each other to varying degrees, therefore it is essential that they should work together, i.e. that they are co-ordinated. For example, muscle contraction is dependent upon respiration, utilising glucose and oxygen; glucose is produced as a result of digestion in the alimentary canal and after absorption is transported to the muscles in the circulatory system; oxygen is supplied via the respiratory system and circulatory system. In addition the activities of the organism are closely linked to changes in the internal and external environment. The organism can detect these changes and reacts in a way which gives it the maximum chance of survival. Co-ordination is an important aspect of **homeostasis** and it is brought about by the nervous system and the endocrine system.

The nervous system

The nervous system provides the quickest means of communication in the body. In most organisms it is divided into a **central nervous system** (C.N.S.) consisting of the brain and spinal cord, and the **peripheral nervous system** consisting of paired cranial nerves (from the brain) and paired spinal nerves (from the spinal cord) which connect receptors and effectors. Each spinal nerve has a dorsal and ventral root. **Receptors** respond to different types of stimulation and thus detect changes in the internal and external environment; they include sense organs such as the eye and ear, and various sensory cells, such as pain receptors in the skin and stretch receptors in the muscles. **Effectors** may be muscles or glands; the information detected by the receptors is conveyed to the C.N.S. via the peripheral nervous system and the appropriate messages are sent to the effectors which respond accordingly.

The basic functional units of the nervous system are nerve cells or **neurones**, the function of which is to receive and transmit impulses ('messages') (their structure is described on page 17). The cell bodies are grouped largely in the C.N.S. and the nerve fibres are arranged in bundles joined by connective tissue to form nerves.

A **synapse** occurs where two neurones meet and transmission of the impulse across the synapse is by chemical transmitter substances, for example acetylcholine.

Reflex action

A reflex action is a rapid, automatic response to a stimulus; it does not require conscious control. In man the knee jerk, coughing, sneezing, blinking, and the removal of the hand from a hot object are a few examples. The nerve pathway between the receptor and the effector involved in a reflex action is known as a **reflex arc**.

A change in the external or internal environment (**a stimulus**) is detected by a receptor, for example temperature receptors in the skin. This results in impulses being transmitted along a **sensory** (afferent) neurone and into the grey matter of the spinal cord via the dorsal root of the spinal nerve; the cell bodies of the sensory neurones are in the dorsal root ganglion. In the grey matter the sensory neurone forms a synapse with an **intermediate** (or relay or association or connector) neurone, which in turn forms a synapse with a **motor** neurone. The axons of the motor neurones leave the spinal cord and transmit impulses to the **effector organ**, either a muscle or a gland. When a hand is removed from a hot object the effector organ is the biceps muscle and this responds by contracting and raising the hand.

Reflexes involving the spinal cord are known as spinal reflexes; those originating from receptors in the head take place through the brain. The brain is not involved in a spinal reflex action but nerve fibres may pass up to the brain to make the person aware of the reflex and any additional behaviour that may be necessary can be carried out.

Conditioned reflexes

In typical reflexes, the stimulus and response are related, for example the taste, smell or sight of food can initiate the salivary reflex. However, it is often possible to replace a normal stimulus by a different stimulus and get the same response; this is known as a conditioned reflex and the organism is said to have been conditioned to the stimulus.

At the beginning of the century a Russian biologist called **Pavlov** carried out his classic work on conditioned reflexes in dogs. In dogs the salivary reflex is initiated by the smell or taste of food and whenever Pavlov fed his dogs he rang a bell. After several days the

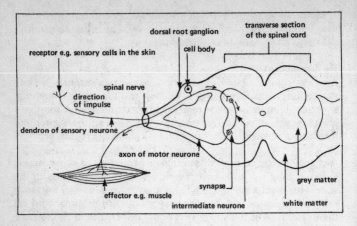

Figure 75. Reflex arc

dogs produced saliva when the bell alone was rung and no food was presented. Thus the normal chemical stimulus of food detected through the nose and tongue had been replaced by an unrelated stimulus detected by the ears. The conditioning can be lost if the bell is rung enough times and food is not presented.

Structure of the C.N.S.

The C.N.S. lies in the mid-line of the body and it consists of the brain and spinal cord. The **spinal cord** is enclosed and protected by the vertebral column, and from it spinal nerves run out between the vertebrae to all parts of the body. On the outside is **white matter** which consists of many nerve fibres running longitudinally to and from the brain and passing out to spinal nerves. Inside this is **grey matter** which in section is seen as roughly H-shaped and it contains many cell bodies. In the centre is the spinal canal which contains nutritive fluid. The spinal cord is very important in spinal reflexes and in transmitting impulses to and from the brain and from one region of the spinal cord to another.

The brain is the enlarged, specialised anterior region of the spinal cord and it has evolved in association with the well-developed sense organs on the head from which it receives sensory fibres. During the course of evolution the brain has developed basically into three regions: the **forebrain** which receives impulses from the nasal organs, the **midbrain** which receives impulses from the eyes, and the **hindbrain** which receives impulses from the skin and ears.

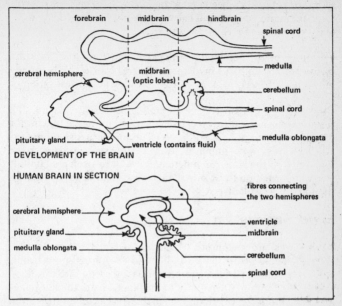

Figure 76. Development and structure of the mammalian brain

In mammals these three regions are not clearly distinguishable due to the great expansion of the roof of the forebrain to form the **cerebral hemispheres** or **cerebrum** which spread backwards over the rest of the brain. The cerebral hemispheres possess sensory, motor and association centres.

The **association centres** contain thousands of nerve cells and receive impulses from the fore, mid and hindbrain. This allows impulses from different sense organs to be sorted out, integrated and the information stored. They enable conditioned reflexes and learning to occur and are responsible for memory, thought, intelligent and complex behaviour patterns, and make us conscious of our activities. The cerebral hemispheres have thus become the centre of integration and co-ordination for the body as a whole and the pattern of behaviour for the whole animal is decided here; it determines the best course of action for the organism to take in a particular situation.

The **cerebellum** is a folded expansion of the hindbrain. It receives sensory fibres from the semi-circular canals, the utriculus, the sacculus and stretch receptors in the muscles. It is very important for maintaining posture and balance. It allows for fine adjustment of movement such as occurs in writing.

The floor of the hindbrain is thickened to form the **medulla oblongata**. It contains involuntary centres controlling breathing movements, heart beat, blood vessels, swallowing and salivation.

Summary The brain receives impulses from sense organs as a result of which impulses are passed along motor nerves to effector organs which respond accordingly. The motor areas and association centres allow impulses from all the sense organs to be integrated so the body can function as a well co-ordinated unit. It allows information to be stored and subsequent behaviour can be modified as a result of it.

Endocrine system

Hormones are chemicals which are secreted by **endocrine glands**. These glands have no ducts so they are also known as **ductless** glands, and their secretions pass directly into the blood circulating through them. Once in the bloodstream the hormones are carried round the body and when they reach certain structures, known as **target organs**, they cause specific changes to take place. Endocrine glands do not function in isolation but influence one another and are integrated into a highly co-ordinated system: the action of the endocrine system provides a further example of **homeostasis**.

A basic similarity between the endocrine system and nervous system is that they are both involved in the co-ordination of the whole organism. They are both set into action by a stimulus to produce a response, and they both involve chemical transmission. Hormones are chemicals and transmission of nerve impulses across synapses is carried out by chemicals, for example acetylcholine, but the distance travelled by the hormones is much further.

There are several basic differences between hormonal and nerve action. Hormones work much more slowly than nerves, and they tend to regulate continuous or long-term processes such as growth or metabolism, whereas nerve action is usually rapid and short-lived, such as when it causes a muscle to contract. Hormonal responses are often widespread, sometimes involving several target organs, whereas nerve responses tend to be more localised, for example they may involve the contraction of one muscle only.

There are many important endocrine glands in the body. The **pituitary gland**, situated on the underside of the brain and linked to it, is sometimes called the 'master' gland because of its effect on other glands. Many pituitary hormones have been isolated and most are proteins. **Growth hormone** influences the growth of bone and other tissues; growth is retarded if the pituitary gland is underactive, and excessive if it is overactive. **ADH** (anti-diuretic hormone) controls the amount of water reabsorbed into the blood in the kidney tubules (see page 124). **FSH** (follicle-stimulating hormone) and **LH** (luteinising hormone) are important in the oestrus cycle (see page 184). In addition the pituitary gland produces hormones which control the production of thyroxine by the thyroid gland, and hormones which regulate blood sugar level and blood pressure.

The **thyroid gland** lies in the neck on either side of the larynx. It secretes the hormone **thyroxine**, which contains iodine, and it controls the metabolic rate. It is therefore important in controlling the rate of growth and development, and undersecretion of thyroxine in infancy causes arrested physical and mental development, a condition known as **cretinism**; in adults it causes a low metabolic rate, accompanied by sluggishness, overweight and coarsening of the skin. Overproduction of thyroxine causes a high metabolic rate accompanied by weight loss, overactivity and an accelerated heart beat. In frogs thyroxine is important in metamorphosis and the addition of thyroxine to the water in which tadpoles are developing promotes metamorphosis.

The two **adrenal glands** lie immediately above each kidney in man, and they consist of an outer cortex and an inner medulla. The medulla secretes the hormone **adrenaline** which has been called the 'flight, fright or fight' hormone; it is released when increased respiration and therefore increased energy production is required to cope with a particular situation. Its effects are not long lasting but its action is similar to that produced by the nervous system supplying the organs of the thorax and abdomen. Adrenaline acts by increasing the metabolic rate, the breathing rate and the heart beat rate; it diverts blood from alimentary canal and skin to the muscles, increases blood sugar level and dilates the pupils.

The **pancreas** lies beneath the stomach and it produces the hormone **insulin**; this hormone controls the level of glucose in the blood by converting excess glucose to glycogen in the liver (see page 74).

The **reproductive organs** produce hormones. The ovary secretes **oestrogens**, and the corpus luteum produced from the ruptured Graafian follicle after ovulation secretes **progesterone**. The function of these hormones is dealt with in detail in chapter 9. **Testosterone** produced by the male in the testes controls the development of the secondary sexual characteristics.

The **stomach** produces the hormone **gastrin** when food enters it and this increases the production of gastric juice. A similar situation occurs in the **duodenum** which secretes **secretin** to stimulate the secretion of digestive juice (see page 70).

The sensory system

An animal must be able to detect changes in its external or internal environment and respond accordingly. These changes or stimuli are detected by **receptors** from which impulses are passed to the C.N.S. The receptors consist of sensory cells which may occur singly, or more or less regularly throughout the organ in which they are found, or they may be concentrated and organised into special sense organs such as the eye. The receptors are connected by nerve fibres to a particular region of the brain. It is only in the brain that the stimulus is identified, according to the region which the impulses entered. A strong stimulus produces a strong response, probably because a greater number of receptors are stimulated.

External receptors respond to stimuli which come from outside the organism, for example, the eye is stimulated by light waves. Internal receptors respond to internal stimuli, for example special sensory cells in the brain are sensitive to the carbon dioxide content and temperature of the blood. In addition, receptors can be classified according to the nature of the stimulus they detect: thus temperature receptors, pain receptors, chemical receptors (taste and smell), light receptors and mechanoreceptors are recognised.

A number of different types of **skin receptors** have been identified. Free nerve endings are sensitive to pain, Meissner's corpuscles under the epidermis are sensitive to touch, while the deeper Pacinian corpuscles are sensitive to pressure. In addition temperature receptors are sensitive to heat and cold.

In the lining of the nasal cavity and on the tongue are **taste cells**. On the tongue they are found in groups called taste buds and there are four kinds sensitive to sweet, bitter, sour and salt. Those at the tip of the tongue are sensitive to sweet substances, those at the back bitter, those at the sides sour, whilst the whole surface seems to be equally sensitive to salt. Taste buds can only detect chemicals in solution so those in dry foods must first dissolve in the moisture of the mouth.

The **olfactory organs** which are sensitive to smell are located in the nasal cavity. For a substance to have a smell it must be volatile, that is it must vaporise, and when it enters the nose it must enter into solution in the film of moisture covering the sensory cells. In man the sense of smell is very easily fatigued.

Pressure receptors in the muscles and tendons respond to the degree of stretching and tension set up in the muscles. This provides information about the position of the limbs and the state of the muscles generally, and this is essential for co-ordinated locomotion.

The eye

The eyes are organs of sight; they are spherical structures which are held in bony sockets or orbits in the skull by six eye muscles. The front of the eye is protected partly by the eyelids and eyelashes. Under the eyelids are tear glands which secrete a fluid containing sodium bicarbonate and sodium chloride; blinking, a reflex action, distributes this fluid across the eye and keeps the conjunctiva moist and washes away dust particles and destroys bacteria. The **conjunctiva** is a thin layer lining the eyelids; it also covers the front of the eye and beneath it in this region is a thick, transparent, protective **cornea** which also refracts light. The **sclerotic coat** is a continuation of the cornea and it forms a thick fibrous protective layer around the eyeball. The cavities of the eye are filled with **aqueous** and **vitreous humours**; they are fluids containing salts, sugars and proteins and they help to refract light and maintain the shape of the eye.

Behind the cornea is the **iris** in the centre of which is a hole, the **pupil**, through which light rays pass to the retina. The iris contains blood vessels, circular and radial muscles, and in brown eyes, pigment. The crystalline **lens** refracts light; it is held in position by **suspensory ligaments** attached to the **ciliary body** which encircles the lens and contains blood vessels and muscles which

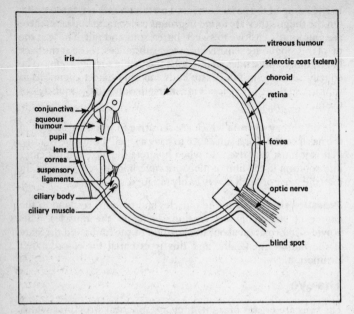

Figure 77. Vertical section through the eye

run in a circular direction. The iris is continuous with the pigmented vascular **choroid** layer which runs round the inside of the eyeball and provides food and oxygen for the light-sensitive retina.

The **retina** lines most of the eye and it contains two kinds of light-sensitive cells called **rods** and **cones**. Rods are more sensitive than cones and function at low light intensities. They contain the pigment rhodopsin or **visual purple** which contains retinene, a substance derived from vitamin A (carotene); lack of vitamin A in the diet can cause night blindness. Cones are sensitive to colour and function in conditions of high light intensity. Nerve fibres run from the base of the rods and cones and they pass across the front of the retina and leave the eye along the optic nerve which goes to the brain. The **blind spot** is the region where the nerve fibres leave the eye and enter the optic nerve; there are no rods and cones here so no image can be formed, but we are not aware of it as the visual fields of our two eyes overlap. The **fovea centralis** or yellow spot is a small depression in the centre of the retina; there are no rods here but the cones are very dense. This is where the image is thrown and it gives detailed appreciation of colour and form.

Formation of an image and vision

Light rays enter the eye and are refracted (bent) by the cornea, lens and humours, and points of light fall on the retina. An image is formed on the retina which is real (on the opposite side from the lens), inverted (upside down) and smaller than the object. The rods and cones are stimulated by the light falling on them and impulses are sent along sensory nerves, down the optic nerve and into the mid-brain where an impression of the form, size and colour of the object is interpreted. The inverted image is corrected in the brain so the object appears upright. The image is usually formed on the fovea.

Control of light intensity

The iris controls the amount of light entering the eye. In bright light the circular muscles of the iris contract, the pupil is reduced in size so less light is admitted and possible damage to the retina is prevented. In dim light, the radial muscles of the iris contract and the circular muscles relax, and this increases the size of the pupil so more light can fall on the retina to increase the brightness of the image. These changes are reflex actions stimulated by a change in light intensity. This explains the gradual ability to see in the dark; it takes a little time for the radial muscles to contract. When passing into bright conditions it is possible to 'see stars', the result of too much light entering the eye.

Accommodation

The lens forms a small, real, inverted image on the retina. So that distant and near objects can be focused, that is form clear images on the retina, it is necessary to change the power of the lens by altering its shape; this is known as accommodation. Near objects require a thick lens with a short focal length and distant objects require a thin lens with a long focal length.

To focus on a near object the circular muscles running round the ciliary body contract; this reduces the diameter of the ciliary body, pulls the sclerotic inwards and reduces the tension on the suspensory ligaments and the lens springs to a thick shape. When the eye is directed towards a distant object the circular muscles in the ciliary body relax, the sclerotic moves out to its original

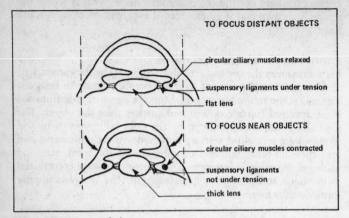

Figure 78. Accommodation

position and the diameter of the eye is increased; this stretches the suspensory ligaments and pulls the lens out to a thin shape. Accommodation is a reflex action and occurs automatically as one glances from near to distant objects.

Stereoscopic vision; judgement of distance

Each eye forms its own image and impulses from the two images are correlated in the brain to give one impression. Each eye sees a slightly different image and the combination of the two allows objects to be seen in three dimensions. The extent to which the eyes converge together with stereoscopic vision gives the impression of distance. Predators such as lions, owls, pike and the apes have eyes set at the front of the head and thus have stereoscopic vision; it is essential for predators to judge the distance of prey before attacking it, but this gives them a limited field of view and they are unable to see behind them. Animals such as rabbits with eyes on the sides of their head form two separate images and they cannot focus clearly or appreciate distance. However, they have a very wide field of vision, and are sensitive to movement of objects against a background; this is known as **parallax**. This is associated with animals that are preyed upon and rely on early detection of their predators.

Defects of vision

Short sight (myopia) is usually caused by the eyeball being too long from front to back. Light from a near object is brought to focus on the retina, but light from distant objects is brought to a focus in front of the retina so the object appears out of focus. It is corrected by wearing glasses with diverging lenses; the rays of light are caused to diverge (bend out) before entering the eye and so are brought to focus on the retina. **Long sight (hypermetropia)** is caused by weak lenses or small eyeballs. Light from a distant object is brought to focus on the retina but light from a near object is brought to focus behind the retina. It is corrected by wearing glasses with converging lenses; the light rays converge (bend inwards) before entering the eye so they are brought to focus on the retina.

Another defect is **lack of accommodation** which is common in old age when the lens loses much of its elasticity and cannot change its shape. **Astigmatism** is caused by irregular curvature of the lens so not all light rays are brought to focus on the retina so some objects are blurred. Colour blindness is more common in men. It is thought that there are three types of cones sensitive to red, blue and green light. **Colour blindness** is an inherited deficiency in cones of one type.

The ear

The ear consists of three parts, an outer, a middle and an inner ear. The outer ear consists of a pinna or external ear flap and an external auditory canal which leads inwards to the **tympanum** or **eardrum**. The **pinna** is attached at the side of the head and is freely movable in most mammals; it is composed of cartilage and directs sound waves into the ear and helps to judge the direction from which they come. The **external auditory canal** contains glands which secrete wax to prevent the entry of foreign particles into the ear.

The middle ear is an air-filled cavity which communicates with the back of the mouth via the Eustachian tube. It contains three **auditory ossicles**, the **malleus** (hammer), the **incus** (anvil), and the **stapes** (stirrup). They are in direct contact with each other and the malleus is in contact with the eardrum and the stapes with the oval window.

The **Eustachian tube** enables the air pressure in the middle ear to remain the same as atmospheric pressure. Atmospheric pressure alters with changes in altitude; when gaining height the air pressure is reduced, so air is released from the middle ear and down

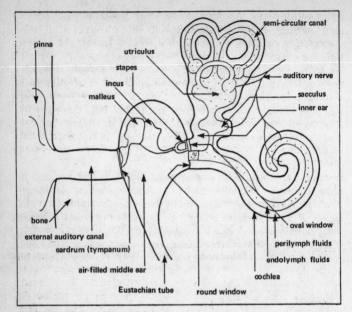

Figure 79. The structure of the ear

the Eustachian tube into the back of the mouth; when losing height the air pressure is increased so more air is admitted into the middle ear from the back of the mouth via the Eustachian tube. This usually occurs without being aware of it but rapid changes in altitude can cause abnormal sensations in the ear.

The inner ear is filled with the fluid perilymph. It contains a coiled tube, the cochlea, semicircular canals, the utriculus and the sacculus. These structures contain the fluid called endolymph and are connected to the brain by branches of the auditory nerve.

The ear performs two basic functions, hearing and balance.

Hearing
The sensory cells which detect sound vibrations are located in the cochlea. In section it is seen to consist of three chambers (canals or ducts). The upper and lower chambers contain perilymph and the former is in contact with the oval window and the latter with the round window. The middle chamber contains endolymph and it is

separated from the upper and lower chambers by membranes. The lower **basilar membrane** consists of transverse stretched fibres running the length of the cochlea and they are sensitive to different wavelengths of sound. Short fibres towards the outer ear respond to high frequency vibrations; these gradually increase in length until at the other end of the cochlea the fibres are long and respond to low frequency vibrations. Above the fibres are **sensory cells**, the upper surfaces of which have hair-like projections which are embedded in the **tectorial membrane** immediately above them. Bases of the sensory cells are connected to branches of the auditory nerve. This part of the cochlea that actually responds to sound is called the **organ of Corti**.

The sound waves or vibrations are collected by the pinna and directed along the external auditory canal to the eardrum which vibrates. The vibrations set up are magnified many times as they are transmitted across the auditory ossicles from the malleus which is in contact with the eardrum, to the incus, and finally to the strapes which is in contact with the oval window. Vibrations of the oval window cause the displacement of fluid in the upper chamber and movement of the upper membrane. This in turn displaces fluid in the middle chamber which moves the basilar membrane thereby displacing fluid in the lower chamber; displacement of this latter fluid is taken up by the stretching of the round window.

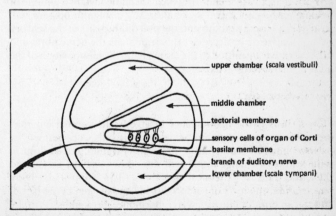

Figure 80. T.S. through the cochlea

How the cochlea responds to sound is still uncertain. Displacement of endolymph in the middle chamber may cause the tectorial membrane to pull the hair cells or perhaps the movement of the basilar membrane moves the sensory cells. However, the sensory cells are distorted by some mechanism and nerve impulses are carried along branches of the auditory nerve to the brain and the pitch, quality and loudness of the sound is interpreted.

The mechanism by which pitch is determined is uncertain. It is probable that groups of stretched fibres only vibrate in response to one particular frequency, so only the sensory hairs resting on these particular fibres would be stimulated; pitch would be determined by the brain detecting which fibres had been stimulated.

Balance

The **semicircular canals** detect movement of the head. At one end of each canal is a swelling, the ampulla, which contains a receptor. Each receptor consists of a group of sensory cells whose hairs are embedded in a gelatinous mass. The semicircular canals are sensitive to movements of the head and as they are arranged in three different planes movement in any direction will be detected. Such movement causes the endolymph in the canals to move towards an ampulla and to displace the gelatinous mass; this stimulates the sensory cells and impulses are passed along the auditory nerve to the brain.

Within the **utriculus** and **sacculus** are receptors consisting of sensory cells, the free ends of which are embedded in a granule of calcium carbonate, called an otolith. Any change in posture will tend to displace the otolith and the pull on the hairs of the sensory cells will increase or decrease; this stimulates the nerve fibres and impulses are transmitted to the brain giving the new position of the head in relation to gravity.

Homeostasis

Homeostasis is the maintenance of a constant internal environment. The chemical reactions which occur in cells work best within narrow limits of temperature, pH, concentration, etc. A drop in temperature or change in pH will slow down enzyme action and thus the rate of chemical reactions. The control of body temperature, the level of carbon dioxide and oxygen in the blood, and the amount of glucose, water, hormones, urea and salts in the blood are all under homeostatic control.

Co-ordination in plants

Plants rely entirely on hormones to co-ordinate the different aspects of their metabolism and to bring about responses to external stimuli. The absence of any equivalent to the nervous system of animals is related to the different methods of feeding. Plants tend to be surrounded by the raw materials they need for photosynthesis and therefore need only to be able to detect and grow towards the most favourable concentration of these materials. Plants show a wide range of movements in response to external stimuli, and in addition other movements which are not directly controlled by external factors. Examples of the latter are the movement of protoplasm within cells, and the movements of guard cells due to changes in turgor pressure.

Movements in response to stimuli in the environment may be conveniently divided into three types. **Nastic** movements are produced by changes in an external factor, but the response to the stimulus is unrelated to the direction of the stimulus. Some common examples of nastic movements are the opening and closing of crocus flowers due to an increase or decrease in temperature, and the opening and closing of daisy flowers due to an increase or decrease in light intensity. **Tactic** movements involve the movement of a whole organism or gamete in response to a stimulus and the response is related to the direction of the stimulus. The plant organisms able to show these movements are unicellular algae with the power of movement. This enables some algae, for example, to move towards areas of favourable light intensity. It is by tactic movements that some male gametes swim to female gametes and bring about fertilisation. The male gametes are responding to a chemical stimulus, and this occurs in algae, mosses and ferns.

Tropic movements are growth curvatures in which the direction of growth is related to the direction of the stimulus. This type of movement is shown by the organs of flowering plants in response to stimuli such as light, water, gravity and touch. The tropic response is named after the stimulus which causes it, so the response to light is called phototropism, the response to gravity geotropism, and the response to water hydrotropism. An organ may show a positive tropism, in which case it grows towards the stimulus, a negative tropism in which case it grows away from the stimulus, or an organ may grow obliquely at an angle to the stimulus.

Main stems, for example, are negatively geotropic (growing away from the pull of gravity), main roots are positively geotropic (growing towards the pull of gravity), and lateral roots grow almost horizontally, at right angles to the pull of gravity. Tropisms may be demonstrated experimentally using seedlings, which are particularly sensitive to the stimuli of light, water and gravity. The coleoptiles of wheat or maize have traditionally been used for investigations into the mechanism of tropic responses.

Experiments to show tropisms

It is possible to demonstrate that stems are positively **phototropic** by placing cress seedlings on moist filter paper in three Petri dishes. One dish is left in uniform light, one dish is left in one-sided light, and one dish is left in the dark. After a few days the seedlings in the light have vertical compact stems and green leaves. The seedlings in one-sided light have stems which bend over towards the light. The seedlings in the dark have long, thin stems and small, pale yellow leaves. These results show that a stem grows towards the light, that is, it is positively phototropic. The seedlings grown in the dark are described as etiolated and show that light is needed for chlorophyll formation, leaf expansion and to retard the rate of growth of the stem.

It is impossible under normal laboratory conditions to eliminate the pull of gravity. In order to demonstrate that roots are positively **geotropic** an instrument called a clinostat is used.

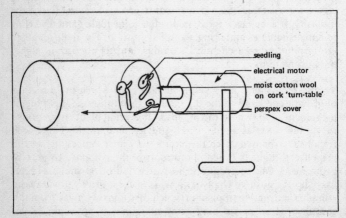

Figure 81. A clinostat

158

This rotates slowly and so subjects all sides of any object fixed to it equally to the pull of gravity. Bean seedlings, with straight radicles, are pinned on to two clinostats in random positions. The seeds are supplied with water by placing them on moist cottonwool and the clinostats are left in the dark for forty-eight hours. One clinostat is stationary and the control clinostat is rotating. The result is that the radicles on the stationary clinostat all grow downwards towards the pull of gravity, whilst those on the rotating clinostat have continued to grow on along their initial direction. This shows that roots are positively geotropic.

It was established by a series of experiments that in a tropic response the stimulus is detected by the tip of the organ and the growth response occurs in the region of elongation immediately behind the tip. This indicates that there must be a link between the two regions, and further experiments showed that this link was a chemical substance. The name **auxin** was given to this chemical before it was identified. An auxin is a plant hormone since it is produced in one region of the plant, transported to another region where it has its effect, and need only be present in minute quantities to produce that effect. The experiments which established these facts were carried out on coleoptiles, and are summarised below.

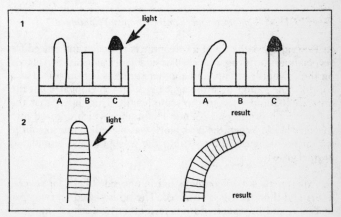

Figure 82. Experiments on the mechanism of tropisms

In Experiment 1 three coleoptiles are subjected to one-sided illumination for several hours. Coleoptile A is normal, coleoptile B has had its tip removed (decapitated) and coleoptile C has its tip covered by foil. The experiment proves that only the tip is sensitive to the light.

In Experiment 2 a coleoptile is marked with lines equal distances apart and then subjected to one-sided illumination. This experiment shows that the region of response to the light is the region of elongation behind the tip.

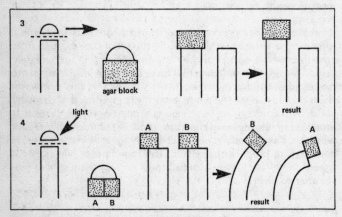

Figure 83. Experiments on the mechanism of tropisms (continued)

In Experiment 3 the tip of a coleoptile is removed and placed for several hours on an agar block. The agar block alone is then placed on the coleoptile stump, and another stump is left untreated as a control. The stump with the agar block on top elongates but the untreated stump remains the same length. This shows that the coleoptile tip produces a chemical substance which diffused into the agar block. When the agar block was placed on the coleoptile stump the chemical diffused into the coleoptile and stimulated normal growth.

In Experiment 4 a coleoptile is left in one-sided light for several hours and then the tip is removed. The tip is placed on two agar blocks, A and B, separated from each other by a strip of material which is impermeable to chemical substances. After several hours the agar blocks are placed on coleoptile stumps which are left in uniform conditions. The coleoptile stump which received the agar block B, from beneath the illuminated side of the coleoptile, bends less. This indicates that the light influences the distribution of auxin so that there is less on the illuminated side of the coleoptile. This causes the illuminated side to grow less than the dark side and so the coleoptile bends towards the light.

These experiments were originally carried out by a Dutch botanist called Went, who also first called the growth hormone involved in the tropic responses auxin. It seems that concentrations of auxin which stimulate growth in shoots inhibit growth in roots. This explains the different responses to gravity by shoots and roots. When a coleoptile tip is placed horizontally and the auxin collected in two separate agar blocks, the lower block receives a higher concentration of auxin than the upper block (see figure 84a). This would lead to the lower half of the coleoptile growing more than the upper half, in normal conditions, and so the shoot would curve upwards, away from the pull of gravity. (Shoots are negatively geotropic.) When a root tip is placed horizontally and the auxin again collected into agar blocks, the lower block had the higher concentration (see figure 84b). In roots this would result in the lower side growing less than the upper side and the root tip turning down towards the pull of gravity. (Roots are positively geotropic.)

It is still not fully understood how the auxin becomes redistributed under the influence of light or gravity, or how the auxin controls the growth rate of cells.

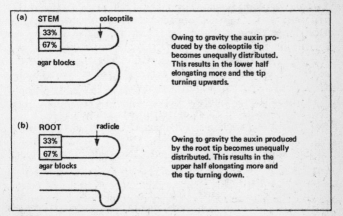

Figure 84. Mechanism of geotropism

It is now known that there are several growth hormones or auxins but the one which has been identified and investigated most fully is called indoleacetic acid (IAA). IAA is believed to be involved in several other aspects of plant growth besides tropic responses. For example IAA inhibits the formation of side branches from lateral buds and stimulates ovaries to develop into fruits. Auxins other than IAA are involved in the control of dormancy, the shedding of leaves and flower formation. There are also many synthetic growth substances which are used in horticulture and agriculture. These substances are used for example as selective weedkillers on lawns, to produce seedless fruits and to promote root formation in cuttings.

Key terms

Accommodation Reflex action whereby the focal length of the lens is altered to enable it to focus near and far objects.

Auxin Plant hormone which influences growth.

Brain Anterior region of the C.N.S. enlarged in connection with aggregation of sense organs in the head; co-ordinates reactions.

Cerebellum Expansion of the hindbrain important in co-ordination of movement and maintaining posture and balance.

Cerebral hemispheres (cerebrum) Expansion of the forebrain associated with memory, intelligent behaviour and consciousness.

Cochlea Part of the inner ear which detects sound vibrations.

Conditioned reflex A reflex in which the normal stimulus is replaced by different stimulus.

Endocrine gland Secretes hormones.

Homeostasis Maintenance of a constant internal environment.

Hormones Chemicals produced in endocrine glands and secreted directly into the blood (which carries them to target organs).

Medulla oblongata Thickening of the floor of the hindbrain; controls involuntary actions such as breathing movements, etc.

Neurone A nerve cell the processes of which conduct impulses.

Receptors Detect changes in the environment.

Reflex action Rapid automatic response to a stimulus.

Semicircular canals Part of inner ear important in balance: detect movement of the head.

Spinal cord Part of the vertebrate C.N.S. within the vertebral column. Important in spinal reflexes.

Stimulus Any change in the environment.

Tropism Growth response directed by one-sided stimulus.

Utriculus Part of inner ear; important in balance, detects changes in posture.

Chapter 8
Growth and Development

Growth is difficult to define satisfactorily since it is often associated with the development of the organism. In basic terms growth is considered as a **permanent increase in size**. This increase usually involves the assimilation of raw materials from the environment, and is accompanied by an increase in dry weight and protoplasm. The synthesis (build-up) of protoplasm requires energy, and this is derived by the process of respiration. Cell division followed by cell expansion and differentiation forms the basis of growth in multicellular organisms. Growth includes some types of plant movements in response to external stimuli; these permanent growth curves are **tropisms**.

Development usually involves a **change in form** and often results in the organism becoming more complex. Growth, unlike development, can be measured; any process which can be measured is a **quantitative process**, e.g. weight or height can be recorded at regular intervals of time. Growth and development usually involve the production of a variety of cells adapted to carry out specific functions. How cell specialisation occurs in cells with identical genes is still not fully understood.

The growth curve

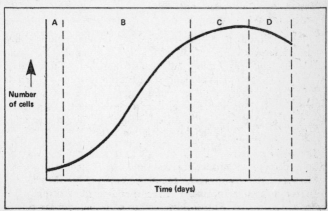

Figure 85. Graph showing growth of unicellular alga

163

Populations, individual organisms and individual organs tend to follow a similar pattern of growth. This may be represented graphically, and although the growth curve is continuous, for convenience it is divided into a number of stages (see figure 85).

A **Lag period** – growth is slow and may not occur at all.

B **Period of rapid growth** – new protoplasm forms at a steady rate.

C **Stationary period** – protoplasm is being produced and broken down at the same rate so there is no overall change in amount.

D **Decline period** – protoplasm is broken down faster than it is replaced; this may be owing to the nutrients being used up or to old age, and it can lead to death.

Growth patterns in animals and plants

In animals growth takes place throughout the body but not all parts grow at the same rate, so animals show changes in proportion as they increase in size. For example, in humans some parts of the body grow at a slower rate than the body as a whole while others grow at a faster rate. As a generalisation, the further away from the brain the faster the growth rate; the brain grows most slowly and stops at about the age of five, while the legs grow at the fastest rate and continue growing until maturity. Thus as a person grows there is a gradual increase in the relative proportions of the legs to the trunk.

In plants growth is usually restricted to certain regions called **meristems**; these are usually at the tips of organs e.g. root and shoot tips, and are therefore called apical meristems. Growth may also occur in other parts of plants such as **cambium**; this is a circle of actively dividing cells in woody stems and roots which is responsible for the increase in girth of these organs. Cell division is followed immediately by rapid cell elongation due to the rapid uptake of water; finally cell differentiation occurs.

Some filamentous **algae**, e.g. *Spirogyra*, are capable of growth by cell division within any cell along their length. Growth may also be induced in normally non-growing regions of plants in response to injury. **Fungi**, like plants, grow only at the tips of their hyphae.

164

Overall growth in many animals e.g. mammals, is continuous until maturity and then stops; this is called **limited growth**. In the adult growth processes may still occur, e.g. the production of new blood cells, but there is no overall increase in size.

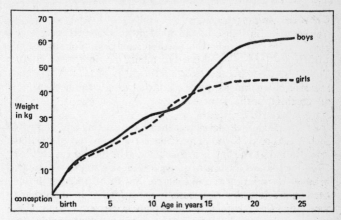

Figure 86. Graph showing growth curve of boys and girls

Most plants retain the ability to increase their size throughout their life; this is called **unlimited growth**. Environmental factors, e.g. seasonal temperature variations, influence growth rate.

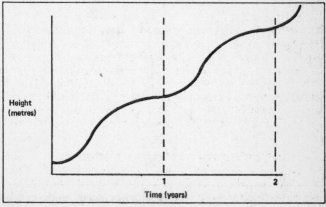

Figure 87. Graph showing growth of a perennial tree, e.g. elm

Growth in the tapeworm, a parasite, resembles that of plants in that there is a continuous production of segments from one region just behind the head. Growth of many marine organisms such as lobsters also tends to be unlimited, and the organisms grow until death. Limited animal growth is a characteristic of land organisms.

Growth in insects

The exoskeleton of insects, and other arthropods such as crustacea, consists of a rigid cuticle which prevents any continuous increase in size. Insects grow by a series of moults or **ecdyses** (singular 'ecdysis'), whereby the existing cuticle is shed and the body increases in volume as the new cuticle expands and hardens; it is during this time that growth occurs.

A new folded cuticle develops beneath the existing one, and when the insect is ready to moult muscular contractions force body fluids into the thorax causing it to swell and split the old cuticle along lines of weakness. Insects may swallow air or water during ecdysis to assist in splitting the old cuticle and to help keep the body expanded while the new cuticle is hardening. In insects with incomplete and complete metamorphosis (see page 173) moulting occurs in the nymphs, larvae or pupae and not in the adults. It is during the moulting period when the cuticle is soft that arthropods are most vulnerable to attack by predators.

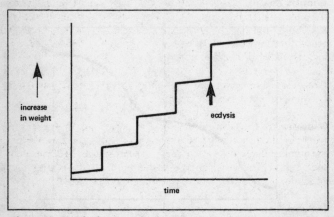

Figure 88. Graph showing changes in weight during growth in an insect

Measurement of growth

Ideally this should involve measuring the increase in protoplasm, but as this is difficult a number of alternative methods are used. The technique employed depends on the organism or organ being studied and the time period involved. With all techniques readings must be taken at **regular** time intervals. If the method selected involves killing the organisms, e.g. dry weight method, valid results may only be obtained by taking large, **random** samples at intervals from a larger population kept under identical conditions. This gives a useful average result.

Methods for organisms and organs

Fresh weight This is an easy technique which does not involve injury to tissues. The whole organism is weighed at intervals. Care must be taken to exclude surface soil or water by some standard method at each weighing. This method does not always give an accurate indication of growth since fresh weight is influenced by fluctuations in the fluid content of the body. It is used to measure growth in vertebrates.

Dry weight This method gives a more meaningful result but involves killing the sample by gently heating it to a constant weight to remove water. It is frequently used for seedlings.

Length This is an easy technique which does not involve killing the material. It is most suitable for organs growing in one direction, e.g. young root tips. Its disadvantage is that it ignores growth in other directions such as width, which may be considerable, e.g. increase in the girth of tree trunks.

Area This has only very limited application for organs growing in two dimensions, e.g. expanding leaves.

Number of organs This method is only applicable to plants where an indication of growth rate may be obtained by counting the number of new organs, e.g. leaves, stems, produced during each time period. The disadvantage of this method is that it ignores variations in organ size, and patterns of growth.

Volume This method is infrequently used as it involves complex mathematical calculations, or immersing the organism in water and measuring the volume of water displaced.

Methods for unicellular organisms

These methods may be used for unicellular organisms such as algae, protozoa and bacteria. These organisms may be grown in solutions, called **cultures**, from which samples may be taken.

The most common technique used is to take many samples of known volume from the culture solution; it is important to have the organisms spread as evenly as possible throughout the culture. The number of organisms in the sample is then calculated by placing the sample liquid on a microscope slide and counting the individual organisms. It is important that a large number of samples is taken each time to obtain a good average reading.

Example calculation of growth in *Chlorella* (a unicellular alga)
The *Chlorella* are grown in a culture solution in a 250 ml flask and kept under constant conditions. As far as possible the *Chlorella* are kept evenly spread throughout the culture solution. 0·1 ml of the solution is withdrawn from the culture solution, examined under the microscope, and the number of *Chlorella* cells counted. A further nine samples of 0·1 ml of solution are removed and the counting procedure is repeated.

Theoretical results for ten readings

Sample	1	2	3	4	5	6	7	8	9	10
No. *Chlorella* per 0·1 ml	98	120	113	124	105	99	118	104	109	113

Calculation

Average number of *Chlorella* in 0·1 ml $= \dfrac{\text{Total number of } Chlorella \text{ counted}}{\text{Total number of samples taken}} = \dfrac{1100}{10}$

$= 110 \ Chlorella/0\cdot1 \ ml \ \text{culture solution}$

Number of *Chlorella* in 250 ml $= 110 \times 2500$

Number of *Chlorella* in 250 ml of culture solution is 275,000.

This method would be repeated at regular time intervals to calculate the rate of growth of *Chlorella*. The results can be expressed graphically.

(Other methods for calculating growth in populations exist but tend to require elaborate laboratory equipment.)

Factors affecting growth

Factors affecting growth do so in many different and subtle ways as demonstrated by the way in which plants respond to daily, seasonal and climatic changes in their environment. The factors influencing growth can be divided into internal and external factors. The overall pattern of the growth of an organism is a result of the interaction of these factors.

Internal factors
This includes the genetic constitution of the organism and the relative quantities of different hormones present in the body. The two are connected in that genes influence growth through the action of hormones.

Genetic material It is the form of chromosomes carried in each cell and it is handed on from generation to generation and gives the organism its potential to grow in a particular way. How this potential is expressed is determined by environmental factors, e.g. genetically identical plants, derived from a clone, grown in exposed and sheltered positions appear very different.

Hormones Plants produce a range of hormones which influence different aspects of growth. These hormones are also essential in co-ordinating the different parts and activities of a plant so it functions well as a whole organism. It was first realised that chemical substances were involved in plant growth during investigations into **phototropism** (the growth of a plant organ in response to one-sided light). The chemical substance was called **auxin**, but it has since been identified as the substance **indole acetic acid** (I.A.A.). Since then other plant hormones have been identified and many are used artificially, e.g. as weedkillers or rooting powder. (See chapter 7.)

Growth in mammals is also regulated by hormones. In mammals the hormone most directly involved is secreted by the **pituitary gland** which lies under the brain; this **growth hormone** is carried in the bloodstream to the bones where it controls protein synthesis. The hormone **thyroxine**, produced from the **thyroid gland** lying in the neck in front of the larynx, controls the rate of metabolism. In young mammals underactivity of the thyroid gland leads to a condition known as **cretinism** where physical and mental development are retarded. (See chapter 7.)

External factors

These include a vast range of environmental influences: light, temperature, the availability of raw materials such as food, water and oxygen, concentration of waste products, physical forces, presence of chemicals, the amount of space, and so on. Some of the most important are discussed below.

Temperature This is very important in controlling the growth of plants and **poikilothermic** ('cold blooded') animals. All chemical reactions in the body, including those involved in growth, are controlled by **enzymes**; temperature has its most significant effect on growth by influencing the rate at which enzymes can act. In general, low temperatures slow down growth and higher temperatures accelerate it, but very high temperatures will inhibit growth. There are exceptional organisms which are adapted to survive in extreme temperatures, e.g. certain bacteria in hot spring water and fish in very cold Antarctic waters.

High temperatures may indirectly slow down growth by making water unavailable.

Plant growth may show a rhythm which is not determined by temperature, e.g. **perennials**, such as lilac, grow most rapidly in spring when the temperature is not the seasonal optimum for enzyme action (see page 29).

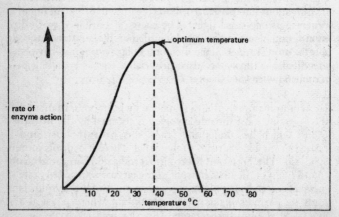

Figure 89. Graph showing rate of enzyme action against temperature

170

Nutrition All organisms must obtain raw materials from their environment for the synthesis of new structures and to provide substances from which the necessary energy is released. If the diet lacks adequate protein, or an essential vitamin or mineral, a **deficiency disease** will occur.

Protein forms the basic structure of the body as two-thirds of a cell is made of protein; a diet deficient in it causes retarded growth and physical weakness in children. The deficiency disease **rickets**, which occurs in children, is caused by a lack of vitamin D; the bones are soft and weak and the legs bow under the weight of the body.

The mineral iodine is important for the production of the hormone thyroxine; a diet deficient in iodine will cause **cretinism** in children. This demonstrates the interaction of external and internal factors in controlling growth; the external factor, iodine, is important for the production of an internal factor, the hormone thyroxine.

Plants absorb, from their surrounding water supply, certain elements which are essential for their growth, e.g. nitrogen (absorbed in the form of nitrates) is essential for the formation of protein, magnesium for the production of chlorophyll. When elements are not present in the correct proportions abnormal growth may occur, as can be demonstrated by 'culture experiments'. In agriculture every attempt is made to provide the correct balance of chemicals in the soil to obtain maximum growth of any particular crop. (See page 118.)

Water This is essential for the growth of all living organisms. Plant cells, unlike animal cells, immediately after cell division take up a large quantity of water which results in a rapid increase in the size of the cell and the formation of the characteristic central vacuole. This can be seen immediately behind the root tip.

The uptake of water and the resulting increase in size or weight is not always considered as a growth change. Only if the water cannot be removed again without damaging the tissues may the gain be defined as growth. For example, the early uptake of water by a seed is not strictly growth since the seed can be dried out again without any apparent damage to it (see chapter 1).

Light This is essential for photosynthesis, the process by which all green plants derive their organic material, and so light is essential for the growth of green plants.

Light is not essential for the germination of seeds since they have a store of organic material at the expense of which initial growth occurs. This accounts for the drop in weight that occurs at the start of seed germination. Seedlings kept in the dark show abnormal growth and are said to be **etiolated**; they have long, thin stems and small leaves which lack chlorophyll.

Plant growth is influenced by light in indirect ways, as can be seen by the difference in the structure of leaves growing in the shade and those on the same plant growing in the light.

Plants exhibit a phenomenon called **photoperiodism** which is the response to the relative length of light and dark periods in any twenty-four hours. This phenomenon influences when some plants will flower, e.g. chrysanthemums will only flower when the light period in twenty-four hours is short, as in England in autumn. This type of plant is called a **short-day** plant. Commercially it is useful to be able to control the flowering (and subsequent fruit production) of plants by controlling the light conditions in which they are grown. In this way it is possible to obtain flowers 'out of season'; chrysanthemums can now be sold throughout the year. Petunias will only flower when the light period in twenty-four hours is long, as in England in summer; this type of plant is called a **long-day** plant. Flowering in some plants is independent of day length, for example in the tomato.

The relative day and night length, along with temperature, influences other aspects of plant growth such as loss of leaves and the opening of buds.

Physical forces Factors such as wind force will influence the pattern of plant growth. This is well illustrated by the distortion of trees on cliff tops caused by strong prevailing winds.

Chemicals Extreme acid or alkaline conditions in the plant's environment may have an adverse effect on its growth.

The drug **thalidomide**, given as a sedative to pregnant women, inhibited the development of the fingers, toes and even limbs of the developing foetuses.

Metamorphosis

Metamorphosis is a change in form during development; it usually involves a series of changes whereby a larva becomes an adult.

Insects have two main types of life history. Insects such as locusts, cockroaches and dragonflies show what is rather misleadingly called **incomplete metamorphosis**. The eggs develop into an adult via a series of **nymphs** which are essentially miniature adults lacking wings and mature reproductive organs. Moulting and growth occurs between each nymphal stage and at each moult the nymph more closely resembles the adult form.

Insects such as butterflies, moths, flies and beetles show **complete metamorphosis**; the egg develops into a **larva** which is strikingly different from the adult. The larvae are called caterpillars, maggots or grubs according to the species. After an active life of feeding and growing when the cuticle may be shed several times, the larva becomes a **pupa** or **chrysalis**. It is during this apparent inactive stage that metamorphosis occurs; there is an extensive breakdown of larval tissues and reorganisation into adult tissues. Eventually the adult or **imago** emerges from the chrysalis.

Metamorphosis also occurs in the **amphibia** when tadpoles becomes frogs or toads. This involves more a modification of pre-existing structures rather than their total replacement; it usually occurs between the tenth and sixteenth weeks after the tadpole hatches from the egg, but the time is partly dependent upon the temperature and composition of the water. The internal gills are replaced by lungs, the legs develop fully, the skin and frilly jaws are shed leaving a wider mouth, the eyes move to the surface of the head, and changes occur to the heart and blood vessels. The tail is then digested and absorbed to provide nourishment for the tadpole which ceases feeding. Finally the young frog climbs out of the pond into damp vegetation.

Significance of metamorphosis

This depends on the particular organisms which undergo metamorphosis. In parasites and sedentary organisms such as sea anemones active larval forms are important to disperse the species.

The numbers of any species are partly controlled by the amount of food and space. In insects with complete metamorphosis and in some which exhibit incomplete metamorphosis the habitat, feeding habits, locomotion and behaviour of the adults are quite different from those of the larvae or nymphs. By exploiting two habitats and two food sources a greater number of individuals of one species can survive, and there is no direct competition for space and food between the feeding and growing larvae and the reproducing adults.

In amphibia metamorphosis is associated with the movement from water to dry land: frogs and toads are imperfectly adapted to life on land; they have external fertilisation and shell-less eggs and thus have to return to water to breed. The stage hatching from the egg, the tadpole, must be aquatic and metamorphosis is essential to change a tadpole with many aquatic features into an adult frog with many terrestrial features.

Metamorphosis in insects and amphibia is controlled by hormones. Injecting a tadpole with the hormone thyroxine can induce metamorphosis producing miniature adults, and similarly lack of iodine (for thyroxine production) in the water can prevent metamorphosis.

Key terms

Assimilation Building up of soluble substances into protoplasm.
Cambium Plant cells capable of cell division.
Development Succession of stages in the life of an organism.
Differentiation Change in the structure or function of cells during development.
Ecdysis Moulting, e.g. when insects shed their cuticle.
Etiolation Condition of seedlings grown in the dark; stems are long, thin and weak, and the leaves are yellow.
Growth A permanent increase in size.
Limited growth Continues until maturity and then stops.
Meristem Localised region of a plant where cell division occurs.
Metamorphosis A change in form during development; usually a series of changes whereby a larva becomes an adult.
Organ Part of a plant or animal composed of several tissues and co-ordinated to carry out a specific function.
Photoperiodism Response of plants to the relative length of day and night.
Quantitative process One which can be measured.
Tissue Group of similar cells with similar functions.
Tropism Growth response of a fixed plant to an external stimulus. Direction of growth is controlled by the direction of the stimulus.
Unlimited growth Continues throughout the life of the organism.

Chapter 9
Reproduction

If a species is to continue it is essential that the organisms within it should produce new individuals which resemble their parents. This process is known as reproduction and it can be carried out **sexually** or **asexually**.

Sexual reproduction usually involves two individuals which produce special sex cells called **gametes**. These fuse at **fertilisation** to form a **zygote**. The zygote then develops into an **embryo** which grows into the adult form. Asexual reproduction does not involve the production or fusion of gametes and it only involves one individual which divides or fragments to produce new individuals.

Asexual reproduction has the advantages of usually being quicker than sexual reproduction, allowing an organism to spread rapidly once it has become established in a suitable environment. The principal disadvantage of asexual reproduction is that the offspring are **genetically identical** and so it does not give the possibility of **variations** which will allow the organism to adapt to changes in the environment. It is more common in plants than animals. Sexual reproduction gives rise to genetically varied offspring and so allows species to withstand changes in an existing environment and to move into new environments.

Sexual reproduction

Sexual reproduction occurs throughout the animal and plant kingdoms from protozoa and algae to mammals and flowering plants. It is now known that even bacteria carry out a form of sexual reproduction. It always involves the fusion of gametes by a process known as fertilisation. In lower organisms the whole living part (the protoplast) of an organism may become transformed into one or more gametes, for example in the alga *Spirogyra*. In addition the gametes may look and behave alike, but as only certain gametes will fuse together in pairs there must be a physiological difference between them and they are referred to as + and − strains: this occurs in *Mucor* and this phenomenon is known as **heterothallism**.

In higher organisms two different sexes, **male** and **female**, are recognised, and certain parts of their bodies have become specialised for the production of gametes. There is a tendency for one gamete to remain stationary, and this is the **egg** or **ovum** produced by the female, and one to be mobile, and this is the **spermatozoon** or **sperm** which is produced by the male; this phenomenon is known as **oogamy**. The ovum is always larger and contains a food reserve; the much smaller sperm cell does not contain a food reserve and moves to the female gamete. In lower plants and animals the sperm cell possesses its own means of locomotion but in flowering plants it is dependent on external agents such as wind or insects for transport.

The number of **chromosomes** is constant for every species and usually each type of chromosome is represented twice, e.g. man has twenty-three pairs of chromosomes and is said to be **diploid**. Each male and female gamete contributes equally to the complete set of chromosomes in an organism and at some stage during sexual reproduction **meiosis** must occur in which the normal body cell chromosome number is halved, i.e. gametes have one set of chromosomes and are said to be **haploid**. Two haploid gametes unite to form a diploid zygote which develops into a diploid organism.

In some organisms both types of gamete are produced within the body of the same organism; they are said to be **hermaphrodite** and this occurs in *Hydra*, the earthworm, the tapeworm, and flowering plants. Some mechanism usually exists to prevent the fusion of gametes from the same organism (**self-fertilisation**). The gametes may ripen at different times, as in *Hydra* and some flowering plants. The structural arrangement of the reproductive organs may make self-fertilisation impossible; this occurs in earthworms and some flowering plants. **Earthworms** copulate in pairs (head to tail pairing) on the surface of the ground on warm, damp nights. It is a complex process which results in an exchange of sperm only; sperm from each worm is passed into special sperm-storing sacs in the other worm. The worms separate and fertilisation occurs sometime later during the formation of the cocoon. Each worm then uses its own eggs and sperm which it received from the other worm during copulation. *Taenia solium* (the pork tapeworm), the adult of which occurs in man, may be up to 5 metres in length and is composed of several hundred segments or proglottides; each segment contains a complete set of male and female reproductive organs and as it is unusual for man (or any other mammal) to possess more than one tapeworm, self fertilisation within one segment, or between adjacent segments, will occur.

Fertilisation may be **external**, occurring outside the body of the female, or **internal** occurring within the body of the female. External fertilisation is common in aquatic organisms where the water provides the medium in which the sperm can swim to the eggs. It also prevents the gametes, zygotes and developing embryos from desiccating, and it allows dispersal. Internal fertilisation is a feature of terrestrial organisms where there is no free water and thus no medium in which the male gametes may swim.

In many organisms sexual reproduction appears to be an adaptation to survival under adverse conditions, such as cold. The fusion product may form a resistant **dormant** stage which only grows into the actual organism when conditions become favourable again, and this gives it time to be dispersed to a new habitat. Examples are the **zygospores** of *Mucor* and *Spirogyra*, the resistant embryo of *Hydra* and the seeds of flowering plants.

During sexual reproduction in insects a resistant pupa is formed. Dormant structures tend to have resistant outer coats and are metabolically inactive, so using up food reserves very slowly. The significance and control of **dormancy** will be discussed in more detail later in the chapter. Figure 108 on page 235 illustrates sexual reproduction in *Mucor* and *Spirogyra*.

Reproduction in vertebrates

Within the vertebrates several trends in reproduction can be recognised: the number of eggs produced is closely related to the type of fertilisation, the amount of protection that the zygote receives, and the amount of parental care given.

In **fish** external fertilisation occurs, the eggs and sperm fusing in the water after they have been shed from the females and males respectively; fish often reproduce in shoals to increase the chances of fertilisation. Most **amphibians** must return to water to breed as they have external fertilisation and shell-less eggs. In both cases the egg is supplied with **yolk** which acts as a food store for the developing embryo until a time when it is able to fend for itself. The mortality rate at all stages is high: the eggs may fail to be fertilised, the fertilised eggs may be eaten or washed away; there is generally no parental care of the eggs or young and as the young hatch at such an immature stage they are eaten in vast numbers. A female cod can lay up to 50 million eggs in a single spawning, and a female frog about 500 to 600 eggs. Fewer eggs are laid in amphibians as fertilisation is more probable; in the breeding season nuptual pads develop on the forelimbs of the male and he uses these to cling to the back of the female during spawning so the male and female are in close proximity. Also the eggs are laid in a clump as frog spawn so they are better protected than the single eggs of fish.

In terrestrial vertebrates two main problems must be overcome if they are to reproduce successfully: as there is no water in which the gametes can swim internal fertilisation must replace external fertilisation, and some way of preventing the developing embryo from desiccating must be evolved. During internal fertilisation the sperm are introduced into the body of the female where fertilisation occurs; fertilisation is more probable and the egg can be enclosed within a protective covering before leaving the body, or it can be retained within the body of the female and develop there with maximum protection.

Reptiles and **birds** have internal fertilisation and the egg is laid in a leathery or chalky shell respectively. The embryo has a supply of yolk and it develops within fluid-filled membranes. Fewer eggs are laid than in amphibia; the young hatch at a later stage of development and are thus more likely to survive. Fewer eggs are laid in birds than in reptiles; in the former the eggs are incubated by the parents and so develop at a constant temperature, and the fledglings are also fed until such a time as they are able to fly and fend for themselves.

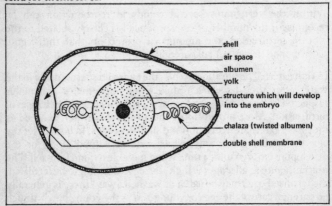

Figure 90. Bird's egg viewed from above, with the top half of the shell removed

In **mammals** few eggs are produced and they are fertilised internally; the male possesses a penis which during mating is introduced into the vagina of the female where the sperm are released. In most mammals the embryo is retained and develops in the **uterus** or womb of the female where it receives a constant supply of food and oxygen. In addition the embryo develops at a constant temperature, it is prevented from desiccating, it is protected against mechanical damage, predators and infection, and the offspring is born at a comparatively advanced stage of development. After birth there is extensive **parental care**. New-born mammals are fed on **milk** produced by the **mammary glands**, and they are fed and protected until they are almost fully grown and able to fend for themselves. In addition parents teach their offspring to cope with various situations in the environment such as hunting for food and avoiding predators. The method of reproduction in the mammal has been a very important factor in their success as a group.

Reproduction in man

Reproduction in man will be described, but the same basic principles apply to all mammals.

The female reproductive organs are called **ovaries**; they are two oval structures lying at the back of the abdomen attached to the body wall. Close to each ovary is the funnel-shaped opening of the **oviduct** or **Fallopian tube** which leads into the muscular

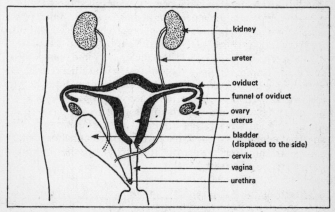

Figure 91. Female reproductive organs: front view

179

uterus or **womb**. The uterus communicates with the outside via the muscular **vagina**; a ring of muscle, the **cervix**, separates the vagina and the uterus.

Between the ages of ten and sixteen the ovaries become active and begin to produce mature eggs; this is the onset of **puberty**. The ovaries begin to secrete **hormones** called **oestrogens** which control **secondary sexual characteristics** such as widening of the hips and development of the breasts. It is thought that there may be up to 400,000 potential eggs or ova in the ovary at birth, but only 200 to 400 will ever mature. Each ovum develops within a structure called a **Graafian follicle**; this consists of a fluid-filled cavity which partly encircles the ovum, nutritive cells, and an outer vascular protective layer. When ripe it is about 12mm long and projects from the surface of the ovary. Finally it bursts and releases the ovum into the funnel of the oviduct; this is known as **ovulation** and it occurs about once every twenty-eight days from alternate ovaries. The ovum is about 0·5mm long and is carried down the oviduct by cilia; it is thought to survive for up to twenty-four hours. After ovulation the Graafian follicle develops into a **corpus luteum** (yellow body), a temporary endocrine organ. If fertilisation does not occur the corpus luteum degenerates; if fertilisation does occur it is retained (page 185). A woman stops ovulating at about the age of forty-five when she reaches the **menopause**.

The male reproductive organs consist of two **testes** which in man lie outside the abdominal cavity in a sac called the **scrotum**; the testes remain at a temperature slightly lower than that of the rest of the body and this favours sperm production. Within each testis are many coiled tubes called **seminiferous tubules** in which sperm are produced. The tubules lead into **vasa efferentia**; these unite to form the **epididymis** which lies outside the testis and which leads into the muscular **sperm duct** or **vas deferens**. Each vas deferens opens into the top of the **urethra** which is prolonged into a **penis**; this contains many blood spaces and at different times carries urine and semen. A short tube, the **seminal vesicle**, branches from each vas deferens before it enters the urethra, and near the urethra at this point are the **Cowper's gland** and the **prostate gland**.

A **spermatozoon** is about 2·5 μm long, which is considerably smaller than the ovum. Men can produce sperm continuously until the age of about seventy.

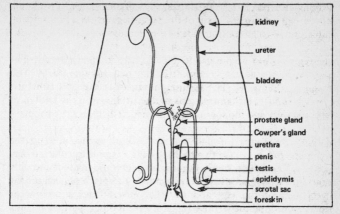

Figure 92. Male reproductive organs: front view

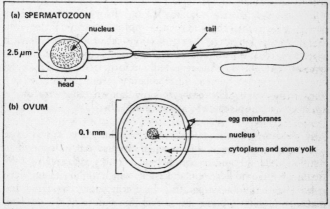

Figure 93. Human (a) spermatozoon, and (b) ovum

Fertilisation

This occurs internally when the sperm meets the egg in the oviduct. For sperm to be introduced into the female it is necessary for the penis to be inserted into the vagina in an act of copulation or coitus. To facilitate this action the penis becomes erect, mainly as a result of blood entering the blood spaces more rapidly than it is leaving and thus increasing the turgidity of the tissues. Stimulation of sensory cells near the tip of the penis triggers off a reflex action causing muscular contractions of the epididymis and vas deferens forcing sperm down the urethra where they are mixed with secretions from the prostate gland and Cowper's gland. The secretions enable the sperm to swim and provide them with essential chemicals. The resulting **semen** is expelled from the penis by muscular contractions of the urethra; this is known as **ejaculation**, and the sperm are deposited at the top of the vagina.

The sperm swim through the cervix and uterus and up the oviducts; this may take two to three days. A single ejaculation may contain 300 million sperm but few will enter the oviduct and only one can fertilise an ovum. The head of the sperm penetrates the cytoplasm of the egg and the two nuclei fuse; a membrane develops around the fertilised egg to prevent other sperm entering.

Pregnancy and development

The zygote undergoes rapid cell division as it passes down the oviduct. This may take up to seven days and by the time it reaches the uterus it consists of a hollow ball of cells which sinks into the lining of the uterus; this is known as **implantation**. As the embryo begins to develop, finger-like projections called **villi** grow from it into the uterus wall and this structure becomes the **placenta**. The cells of the embryo divide repeatedly to form tissues and many of the organ systems, such as the circulatory system, are formed at a very early stage. The embryo, now known as a **foetus**, is connected to the placenta by an **umbilical cord**.

The placenta is a large disc of tissue which develops in close association with the highly vascular uterus wall. It contains a network of blood capillaries and blood is conveyed to and from the foetus by the umbilical vein and artery which run in the umbilical cord. The membranes separating the uterine blood capillaries and placental blood capillaries are so thin that diffusion of substances between the capillaries occurs. Dissolved oxygen, glucose, amino-acids, salts and vitamins diffuse from the uterine capillaries across into the placental capillaries and are carried down the umbilical vein to the foetus. Dissolved carbon dioxide and excretory products, such as urea, from the foetus pass along the umbilical artery into the placenta and diffuse into the uterine blood capillaries. The placenta can act as a selective barrier to prevent the entry of unwanted substances from the maternal blood. At no time do the maternal and foetal circulations come into direct contact.

The villi give the uterus and placenta a greater surface area for the exchange of substances.

The foetus is enclosed in a double membrane, the **water sac**; the inner membrane, the **amnion**, encloses the fluid-filled **amniotic cavity** in which the foetus is suspended. The **amniotic fluid** cushions and protects the developing foetus.

To accommodate the growing foetus and amniotic cavity the uterus enlarges considerably. At five months the foetus begins to move its limbs and at about nine months it is ready to be born. The time from fertilisation to birth is known as the **gestation period**. Before birth the head of the foetus comes to lie just above the cervix.

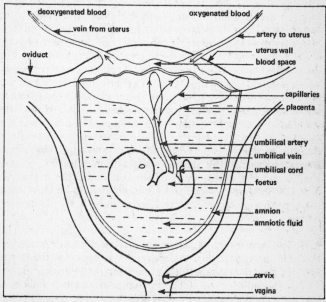

Figure 94. Relationship between foetus, uterus and placenta

In the process of birth (**parturition**) the muscular walls of the uterus begin to contract rhythmically; this is the onset of labour and the contractions become more frequent. The cervix dilates to enable the baby's head to pass through and at some stage the water sac bursts. Muscular contractions of the uterus and abdomen eventually expel the baby head first through the dilated cervix and vagina. The sudden drop in temperature stimulates the newborn baby to breathe and this is usually accompanied by crying. The umbilical cord is tied and cut and where the cord comes away from the baby a scar called the navel remains. A little later the placenta comes away from the uterus wall and it is expelled as the **afterbirth**. The baby is suckled on milk which is often regarded as an **'ideal food'**: it contains all the carbohydrates (lactose), fats, proteins, vitamins and mineral salts that a baby needs for growth and respiration; it lacks vitamin C and iron, but the baby has an iron store which was built up in the liver during pregnancy. Parental care in humans is the most lengthy of any mammal.

Hormones involved in the menstrual cycle and pregnancy
The reproductive cycle in females is known as the **menstrual** or **oestrus cycle**. It lasts about twenty-eight days and follows a set pattern which is regulated by hormones secreted from the **pituitary gland** and **ovary**. If pregnancy occurs the normal cycle is interrupted and in addition the placenta produces hormones. On day 1 of the 28-day cycle menstruation begins as a result of events which occurred during the previous twenty-eight days. Also at the beginning of the cycle the pituitary gland, which is situated under the brain, is stimulated to produce the hormone **F.S.H.** (follicle stimulating hormone). F.S.H. is carried in the bloodstream to the ovary where it causes Graafian follicles to develop. As a result of this the ovary secretes the hormone **oestrogen** which has several effects: it repairs the uterus wall, and it is carried in the bloodstream to the pituitary gland where it inhibits the production of F.S.H. so no more Graafian follicles develop, and it also causes the pituitary gland to secrete the hormone **L.H.** (luteinising hormone).

L.H. is carried in the bloodstream to the ovary where it causes ovulation on day 10 to 14 of the cycle; at ovulation the egg is released from the Graafian follicle and is carried down the oviduct. Under the influence of L.H. the ruptured Graafian follicle is converted into a corpus luteum (yellow body). The corpus luteum becomes a temporary endocrine organ and begins to secrete the hormone **progesterone**. Progesterone also inhibits F.S.H. production so no more Graafian follicles develop and it builds up and prepares the uterus lining ready for implantation of a fertilised egg.

If fertilisation does not occur the corpus luteum begins to

degenerate towards the end of the 28-day cycle and the production of progesterone declines; this has two effects which occur in the next cycle: the lining of the uterus (the endometrium) comes away and is discharged through the vagina; this is known as **menstruation** and it occurs on day 1 of the next cycle and lasts for about five to seven days. F.S.H. production is no longer inhibited so F.S.H. is secreted and Graafian follicle development begins again.

Pregnancy If fertilisation occurs, the fertilised egg becomes implanted in the uterus wall and the corpus luteum does not degenerate. It continues to secrete progesterone which maintains the uterus lining and causes the mammary glands to enlarge. Later in pregnancy the placenta takes over the job of secreting progesterone and oestrogen. At parturition (birth) the level of progesterone declines, allowing oestrogen and another hormone from the pituitary gland, **oxytocin**, to take effect and cause contractions of the uterus which expel the baby through the cervix and vagina. Suckling causes another pituitary gland hormone to be released and this causes secretion of milk from the mammary glands. After birth a decrease in progesterone and oestrogen occurs and this causes F.S.H. to be released again. Two hormones inhibit F.S.H. production and therefore Graafian follicle development. These are progesterone and oestrogen, the two hormones which form the basis of the contraceptive pill.

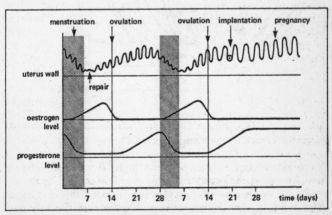

Figure 95. Relationship between uterus wall and progesterone levels

Reproduction in angiosperms

The most advanced land plants are the angiosperms (flowering plants) and their process of **sexual** reproduction deals efficiently with the problems of stationary individuals, no free water for the transference of gametes, and dispersal of the embryo. The flower contains the organs of sexual reproduction, and a generalised flower cut in half appears in figure 96. Most species have flowers which contain both male and female organs (e.g. rose and buttercup), and a few species have flowers which contain only male or female organs (e.g. hazel).

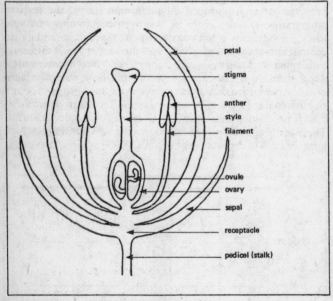

Figure 96. Vertical section through flower

The organs of the flower are borne on the swollen end of the flower stalk known as the **receptacle**. In some flowers (e.g. apple) the ovary is actually embedded in the receptacle and this is called an inferior ovary. The outermost organs, which are the only ones visible in the flower bud, are the sepals, the entire whorl of sepals making up the **calyx**. The function of the sepals is generally to protect the flower in the bud stages, and they are usually green and tough in texture. The next whorl of organs, called the **corolla**, is made up of the petals. These are generally delicate, brightly coloured, may be fragrantly scented, and produce nectar at their base. They serve as organs of attraction for insect visitors. Inside the petal is a whorl of stamens which makes up the male reproductive organ called the **androecium**. Each stamen consists of an anther-head borne on a slender stalk or filament. The innermost structures are the female reproductive organs, or carpels. The carpels make up the **gynaecium**. Each carpel consists of a swollen ovary containing one or many ovules, and a receptive surface, or stigma, borne up on a stalk or style. A collection of flowers on a single main stalk is called an **inflorescence** (e.g. bluebell). Flowers which are symmetrical and may be cut vertically in any plane to produce two identical halves are called **actinomorphic** (e.g. buttercup). Flowers which are asymmetrical and may be cut in only one vertical plane to produce two equal halves are called **zygomorphic** (e.g. sweet pea). In the family of flowers called the Compositae, a large number of small flowers are grouped together on a single receptacle (e.g. dandelion). An example of each of these three types of flower (actinomorphic, zygomorphic and Compositae) should be examined practically.

The function of the **stamens** is to produce the male gametes. Each stamen consists of four pollen-sacs in which the pollen grains grow.

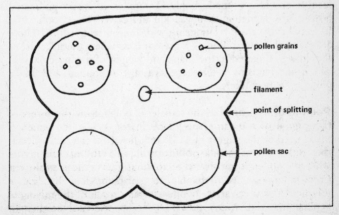

Figure 97. T.S. through anther

The pollen grains contain the male gamete and so must be transferred to the female reproductive organ if fertilisation is to occur. The pollen sacs split along definite lines to release the pollen. The transfer of pollen may be brought about by insects or wind. Many features of the flower are influenced by which agency is used for pollination. In wind pollination it is purely chance whether the pollen released from the anthers of one flower reaches the stigma of another flower of the same species. For this reason a large quantity of pollen must be produced, as only a small percentage is effective in bringing about pollination. The pollen is light and dry so that it is easily carried by the wind. Wind-pollinated plants, such as grasses and hazel, have anthers which hang out of the flower on long filaments, so that the wind is more likely to dislodge the pollen. The stigmas of wind-pollinated plants are feathery and outside the flower so as to expose a large surface area to trap pollen grains. Wind-pollinated plants produce small, inconspicuous flowers without any scent or nectar. This is because the flowers have no need to attract insects.

In insect pollination the pollen is more likely to reach the stigma of a flower of the same species and so a smaller quantity of pollen is produced. The pollen is larger and sticky so that it adheres to the body of the insect. The anthers and stigma are protected inside the flower in a position where the insect is likely to touch them when entering the flower. The flowers of insect-pollinated plants attract insect visitors by producing nectar and scent, and by being colourful. The latter two attract the insects' attention and the former provides a food supply. Examples of insect-pollinated flowers are dandelion, honeysuckle and clover. Insect-pollinated flowers have a close relationship with their insect visitors. There are many elaborate pollination mechanisms where the structure of the flower is related to the shape and weight of the insect visitor. These mechanisms reduce the wastage of pollen and make pollination more likely.

Pollination is defined as the transfer of pollen from the anther to the stigma of the same species of flower. When the pollen is transferred to the stigma of a different flower of the same species this is known as **cross-pollination**; **self-pollination** occurs when the pollen is transferred on to the stigma of the same flower. Cross-pollination is preferable to self-pollination because it produces a greater variety of offspring as a result of the mixing of genetic material. Many flowers have specific mechanisms for reducing the chance of self-pollination and making cross-pollination more likely. The commonest method is for the anthers and stigma in the same flower not to ripen at the same time, for example in dandelion flowers the stamens ripen before the stigma.

The **carpel** is more complex in structure and function than the stamen. Each carpel must have a receptive surface, the stigma, to capture pollen grains. The ovule contains the female gamete which after fertilisation develops into a seed containing an embryo. The ovary protects the ovule and after fertilisation develops into a fruit which is important in dispersing or scattering the seeds.

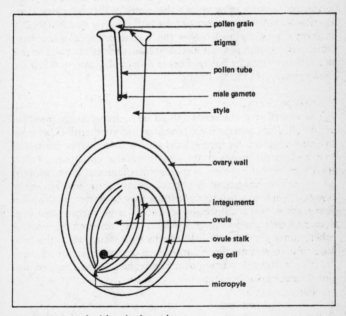

Figure 98. Carpel with a single ovule

Fertilisation

Once a pollen grain has been deposited on the stigma, pollination has occurred. The pollen grain will only start to grow if it is placed on a ripe stigma of a flower of the same species. The stigma produces a sticky, sugary liquid when ripe to allow pollen grains to germinate. Each pollen grain produces a pollen tube which pushes its way down between the cells of the style until it reaches

189

the ovule. When this occurs the male gamete, which has travelled down the pollen tube, enters the ovule through the micropyle at its tip and fuses with the egg cell or ovum. Fertilisation has occurred, and this might be hours or weeks after pollination occurred, depending on the species. Each ovule is borne on a stalk attached to the ovary wall. The ovule is almost completely covered by two outer layers called integuments which only leave a small gap on the surface of the ovule, called the micropyle. After fertilisation the egg cell develops into the embryo, the ovule develops into the seed, and the ovary develops into the fruit. The embryo rapidly forms a potential stem, root and leaves. The ovule drops its water content down to about 10 per cent and the integuments become a tough protective seed coat or testa. The ovary wall undergoes large changes to form the pericarp or fruit wall. This varies considerably in form depending on whether the species produces a dry hard fruit, such as a nut, or a soft succulent fruit, such as a berry. The rest of the flower usually has no further role to play after pollination, and petals, sepals, etc. shrivel up.

Dispersal

It is essential that the seeds should be scattered away from the parent plant to prevent overcrowding and competition between members of the same species. Dispersal is aided by the dryness of the seed which means it can survive adverse conditions, and is light. A large range of dispersal mechanisms have evolved involving the shedding of individual seeds or entire fruits from the parent plant. Light, dry seeds or fruits may be carried by the wind using wings or parachutes to produce a large surface area (e.g. sycamore and dandelion). Juicy fruits may be eaten by birds or other animals and the seeds inside the fruits scattered in this way (e.g. tomato and mistletoe). Ingenious devices of shooting seeds out by the sudden splitting of the fruits help to get the seed into more turbulent air (e.g. lupin).

Seeds and germination

Seeds vary considerably in detail both in their structure and pattern of germination. They do, nevertheless, have a number of features in common. The embryo always consists of one or two seed leaves or cotyledons, a stem or plumule, and a root or radicle. Plants which contain one cotyledon in their seed are called monocotyledons, and plants with two cotyledons are called dicotyledons. These two groups of plants differ in several other respects also. Examples of the former are grasses and palms, and examples of the latter are shrubs and the majority of deciduous trees. In some seeds (e.g.

broad bean and pea) the cotyledons are large and fleshy since they are swollen with most of the seed's food reserves. In the case of the examples given, the reserves are mainly in the form of starch. Other seeds store most of the food outside the embryo in a special region called the endosperm (e.g. maize). The testa or seed coat is a protective layer reducing the damage from frost and the risk of the food reserves being eaten or rotting away. The micropyle remains as a minute hole or pore in the testa and allows the rapid entry of water at the onset of germination.

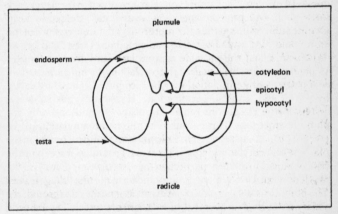

Figure 99. Structure of a seed

All seeds require a basic set of conditions to germinate, which may be demonstrated experimentally. These conditions are an adequate supply of water, oxygen and a suitable temperature. Many seeds when released from the parent plant and placed in favourable conditions still fail to germinate. These seeds are described as **dormant** and need some specific treatment or time period before they are capable of germinating. Examples of factors causing dormancy are that the embryo may still be immature, the testa is acting as a mechanical strait-jacket, or a chemical inhibitor is present in the seed preventing germination. Dormancy is important because it allows time for the seed to be dispersed and makes it more likely to start germination when conditions are suitable in its particular environment. For example, some desert plants contain a chemical inhibitor and require a heavy rainfall to wash all the inhibitor away. This means that the seed will only germinate when there is sufficient water to allow the seedling to become well established. Seeds may remain dormant from only a few days up to a thousand years, depending on the species.

Germination is the initial stages of growth of the embryo. It involves the rapid uptake of water, changing the water content of the seed from 10 per cent back up to about 90 per cent. Providing that the temperature is suitable, the water allows enzymes to convert the insoluble stored food reserves into a soluble form and for these soluble products to reach the plumule and radicle. The plumule and radicle then start to grow. During germination there is a rapid uptake of oxygen as sugars are being broken down to release energy for growth. Much of the energy is wasted and released as heat. Germination results in the testa rupturing and the emergence of the embryonic root and shoot. The radicle appears first, and grows downwards, and the plumule appears later and grows upwards. Depending on the behaviour of the cotyledons during germination, two patterns of germination may be recognised. These are called hypogeal and epigeal germination. **Hypogeal germination** results in the cotyledons remaining below ground and only the plumule emerging. This is caused by the rapid elongation of the epicotyl (e.g. broad bean seed). **Epigeal germination** results in the plumule and cotyledons appearing above ground as a result of the rapid elongation of the hypocotyl. In these seeds (e.g. sunflower) the cotyledons turn green above ground and form the first foliage leaves. The plumule is delicate and to protect it as it pushes up through the soil it may be hooked over (e.g. broad bean) or protected by a sheath, called the coleoptile, as in maize and wheat. The radicle is protected by a root cap.

Asexual reproduction

This process is common throughout the plant kingdom but only occurs in some simple animals. Bacteria, some protozoa and algae demonstrate the most simple form of asexual reproduction, that of **binary fission**. This involves the organism dividing into two equal-sized parts, and the process can lead to a rapid increase in the number of individuals. For example, a bacterium under favourable conditions may divide once every twenty minutes. Many plants, and some protozoa, produce spores. These are usually produced in large numbers and are small unicellular structures. They are light so that they are easily dispersed by water and wind or animals. The airborne species of bacteria and fungi are well known to man owing to the damage they may cause to crops, livestock or himself.

Mucor, the common bread mould, produces spores asexually as a means of spreading rapidly from one food source to another. This is important for saprophytes since their food supply may only exist for a limited period. A large number of spores is produced since so many are wasted, landing on unfavourable substrates.

Some lower animals, for example *Hydra*, and the fungus yeast, can produce asexually by a process of **budding**. A daughter organism develops on the side of the parent, and then having reached a suitably advanced stage becomes detached. Yeast is an unusual fungus in having rounded cells and not tubular hyphae. These cells are naturally present on fruits and man has benefited from their activities in brewing and wine-making. In favourable conditions a yeast cell may bud so rapidly as to produce chains of cells.

Vegetative propagation is a very common method of asexual reproduction in flowering plants. This occurs when part of the parent plant structure becomes detached and forms a new individual. This type of reproduction tends to be characteristic of herbaceous perennials. These are plants which live for several years but at the end of each growing season leave no persistent aerial system. Over winter a storage organ remains below ground which allows rapid growth, as soon as conditions are favourable, next spring. One or several individuals may be produced from each storage organ. These organs which allow the plant to survive during adverse conditions, as well as carry out asexual reproduction, are called organs of **perennation**. These organs vary from species to species, depending on which region of the plant stores the food, but are alike in that they all consist of a bud or buds borne on a food reserve. The potato tuber is a swollen end of an underground stem, the onion bulb is the swollen base of foliage leaves and the carrot is a swollen tap root. Each year the foliage leaves of these plants send down to the storage organ excess foods made in photosynthesis. These foods are translocated in a soluble form, such as sugars, and then converted to an insoluble form, such as starch, by enzymes in the storage organ. The following spring these reserves are converted back into a soluble form and used for growth and respiration by the buds.

Organs which carry out vegetative propagation are particularly valuable in that they allow a rapid spread into a new environment, once the parent plant has become initially established. An example of this type of organ is the strawberry runner, which is a side branch sent out from the parent plant. New plants form at the nodes of the runner and then become separated when the internodal region of the runner decays.

Man has developed artificial means of vegetative propagation by inducing buds to form new plants. The two methods used are grafts and cuttings. **Cuttings** are short lengths of shoot made to form adventitious roots more readily by the use of 'rooting powders', which are plant hormones. Pansies and geraniums are commonly propagated by means of cuttings. **Grafts** may be made in a variety of ways and are used most extensively in the cultivation of rose and fruit trees. The variety of fruit or flower required, called the scion, is attached in the form of a twig to a well-established root system, called the stock. It is essential that the cambium of the scion and stock are in contact and that the join between them is bound to prevent infection and frost damage.

Key terms

Asexual reproduction Reproduction without gametes, and it can only involve one individual.
Dormancy A resting stage in which metabolism is very slow.
Epigeal Seed germination in which the cotyledons are carried above the ground, e.g. sunflower.
Fertilisation The fusion of two gametes during sexual reproduction to form a zygote.
Flower Collection of reproductive organs together with organs of protection and attraction.
Fruit Develops from the ovary after fertilisation.
Gamete Reproductive cell produced during sexual reproduction.
Germination Growth of embryo of a seed into an independent plant.
Hypogeal Seed germination in which the cotyledons remain below the ground, e.g. broad bean.
Ovum Egg cell; female gamete before fertilisation.
Placenta Structure formed between the embryo and uterus wall in most mammals; it allows the exchange of substances between the maternal and the embryonic blood.
Pollination Transfer of pollen from the anther to a stigma of the same species of plant.
Seed A fertilised ovule which contains an embryo.
Sexual reproduction Reproduction with gametes, usually involving two individuals.
Sperm A spermatozoon; male gamete.
Zygote Structure formed at fertilisation when the gametes fuse.

Chapter 10
Genetics and Evolution

Genetics, or heredity, is the science which studies the resemblances and differences between parents and their offspring. Every species has a set of characteristics which all the members share in common. These characteristics distinguish one species from another. This type of variation between different groups of organisms is the basis on which the classification system works. Even within a species individuals still vary; the different races of man, all members of the same species, illustrate this point clearly. Also an individual varies during the course of its development. There are two possible causes of these variations, the genetic material of the individual (called its **genotype**), or environmental factors. Differences caused by the genetic material are inheritable (that is, handed on from generation to generation), but differences produced by the environment are non-heritable. Some human examples of heritable characteristics are eye colour, finger prints and blood groups. None of these is controlled by the environment. Eye colour in man is slightly different from blood groups, however, in that there is a continuous range of eye colours from blue, through green, hazel, etc., to brown. On the other hand there are only a limited number of quite separate blood groups into which humans may be placed. Eye colour is said to be an example of continuous variation and blood groups an example of discontinuous variation.

Some environmental factors which may influence the observable features of an organism are the amount and type of food available, chemicals present and the temperature. The fact that the environment of an unborn child contained thalidomide produced horrifying effects on the baby's development. This resulted in babies being born without toes or arms. It is apparent that the organism's genotype and environment may interact to produce its **phenotype** (all its observable features, both internal and external). For example, man genetically has the potential of reaching a certain height, but this may only be fully expressed providing that environmental factors, such as available food, are favourable.

The laws of heredity were first proposed by an Austrian monk, **Gregor Mendel**, working in 1856 to 1865. During this time he carried out a great many breeding experiments using the garden pea, *Pisum sativum*, and it was his choice of material which helped so much in his success. Mendel's first experiments involved crossing plants which differed in one characteristic only. All his plants were true breeding, which means all the offspring resembled their parents. For example, he crossed a pure breeding tall plant with a pure breeding dwarf plant. This cross, and the result from growing the seeds produced, may be represented as follows: P stands for parental generation, F_1 for the first generation of offspring, and the next generation would be represented by F_2, etc.

P   tall x dwarf

F_1 tall (self-fertilised)

F_2 787 tall   277 dwarf

The garden pea naturally self-fertilises and Mendel had to transfer pollen from one flower to another when he cross-fertilised the parent plants. Mendel took care to grow the plants under the same environmental conditions and in all his crosses he investigated whether it made any difference if the male or female carried the character. He carried out many similar experiments from which he drew a number of observations. In all cases the F_1 resembled one of the parents, in the example used the F_1 resembled the tall parent. This character Mendel called **dominant** and its alternative, the **recessive** character. In the F_2 Mendel found that both characters were present, in numbers always close to the ratio of 3 dominant to 1 recessive. Although Mendel knew nothing about the existence of chromosomes or genes, he suggested that the characters were handed from generation to generation through the gametes. In the F_1 the characters did not mix or blend but remained separate, one character (the recessive) merely being hidden by the other character (the dominant). In the F_2 both characters could appear unaltered.

It was not until 1900 that the term **gene** was first used. A gene is a small piece of genetic information which can exist in one of two forms. These two forms of a gene produce the two alternative forms of the same character (e.g. short and tall are alternative forms of the character height). The alternative forms of a gene are called **alleles**. The allele which produces the dominant feature is always represented by the capital letter, and the allele producing the recessive feature by the small letter (e.g. the dominant allele producing tallness by T, and the recessive allele producing dwarf by t). Now that we know more about the structure and behaviour of chromosomes, we can represent Mendel's experiments in a different way. Diploid organisms contain two sets of chromosomes, one set obtained from the female's egg cell, and one set obtained from the male's sperm. As a result of meiosis each gamete contains only one allele from any pair of alleles. Therefore each gene (carried on the chromosomes) will be represented twice. In an organism which is pure breeding for a certain character both alleles controlling that character will be the same, for example TT (pure breeding tall) or tt (pure breeding dwarf). This organism is said to be **homozygous** for that character. An organism which is not pure breeding for a character contains both alleles, for example Tt (not true breeding tall). The organism is then said to be **heterozygous** for that character.

Mendel's cross between tall and dwarf plants may be represented by figure 100.

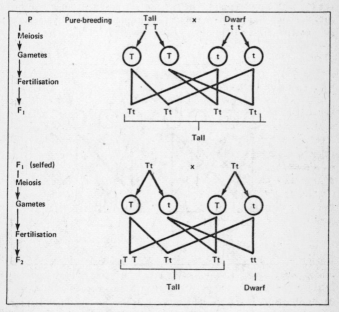

Figure 100. Inheritance of height in Mendel's peas

In order to obtain Mendel's 3:1 ratio a large number of crosses must be carried out because the result is determined by chance. Just as in the same way a coin must be tossed many times to get 50 per cent heads and 50 per cent tails! Also the biological assumptions are made that the heterozygote (Tt) produces the two possible types of gametes T or t in equal proportions, and it is purely chance that controls which male and female gametes fuse. Mendel was fortunate in that these two situations applied in all his experiments; there are some organisms which produce exceptions to these rules and would have prevented him from reaching his important general conclusions. Also Mendel did carry out each cross many times so that the results could be mathematically analysed.

In breeding experiments it is often important to be able to distinguish between the double dominant homozygotes and heterozygotes. These have the same phenotype (appearance) but different genotypes. A recessive **back cross** is carried out to determine which is the true breeding plant, and involves crossing the tall with a dwarf plant. The two possible results from crossing a homozygous and a heterozygous tall pea plant with a dwarf plant are shown in figure 101.

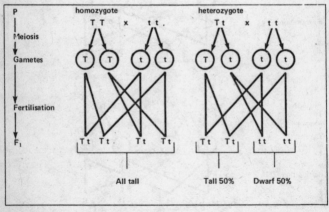

Figure 101. Recessive back cross

We said earlier that Mendel was wise in his choice of material for his breeding experiments. We now know of many exceptions which do not produce such clear-cut results, and would not have enabled Mendel to produce his basic laws of genetics. One such exception is known as **incomplete dominance**; this occurs when within a pair of alleles neither is dominant to the other. An example of this is the inheritance of flower colour in the *Antirrhinum* (snapdragon). In figure 102 R represents the allele producing red flowers and r the allele producing white flowers. The results of crossing a red and white flowered plant are shown in the figure.

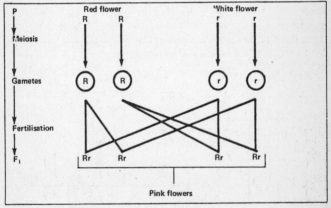

Figure 102. Example of incomplete dominance

In this case each genotype has a different phenotype since there is no masking effect in the heterozygote. RR is red, Rr is pink and rr is white.

Chromosomes and genes

We said earlier that chromosomes contain the genetic material, carrying the instructions needed by the cell to carry out its normal functions. We have also said that genes control the phenotype of an organism. What then is the relationship between chromosomes and genes, and genes and cell activity? It has been shown by many breeding experiments that genes control the production of enzymes. Since genes only occur in the nucleus, and enzymes are made on the ribosomes in the cytoplasm, there must be a

199

messenger which carries the information from the nucleus to the cytoplasm. This messenger is ribonucleic acid (RNA). The method by which the genetic material controls the synthesis of proteins is explained simply below.

A sequence of three particular bases along the length of the deoxyribonucleic acid (DNA) molecule corresponds to one particular amino-acid. The DNA makes a copy of these base sequences in the form of messenger RNA, and the messenger RNA then moves to the ribosomes. At the ribosomes the sequence of bases on the RNA controls the order in which amino-acids are joined together. One gene is thought to control the synthesis of one protein. This may be represented diagramatically as in figure 103.

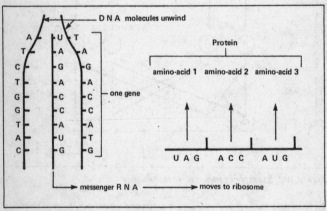

Figure 103. DNA control of protein synthesis

The messenger RNA moves through the pore in the nuclear membrane to reach the ribosomes on the endoplasmic reticulum. Energy is required to join the amino-acids up to form a protein. Since all enzymes are proteins, the DNA in this way controls the **metabolism** of the cell.

Mutations

These are random unpredictable processes. They may be of two kinds, chromosome mutations involving some large structural change in the chromosomes, and gene mutations involving a chemical change in an individual gene. Mutations occur within genetic material at a certain natural rate, but this rate may be increased by certain factors in the environment. Some examples of the factors are X-rays, ultra-violet light, radioactivity and certain chemicals such as formaldehyde. Most mutations are harmful to the organism since they interfere with a carefully balanced system. The most significant place for a mutation to occur is in a gamete, or zygote, since it will then influence a complete organism. This is why extra care is employed when taking X-rays in regions close to the reproductive organs, or in pregnant women. Mutations in body cells produce a 'mosaic effect' because just a spot or patch on the organism is different. Although mutation rates are very low – perhaps one gene will only change one millionth of the individuals in each generation – it is the source of many of the differences inherited within a species. Mutation results in the formation of new alleles, or alternative forms of gene. The new allele produced may be either dominant or recessive, but the majority of mutant alleles are recessive. A recessive mutant allele in a heterozygote will not show immediately as its effect will be masked in the normal dominant allele.

Sex determination

The method by which the sex of an individual is determined varies depending on the species. In man it is controlled by a single pair of chromosomes, called the X and Y chromosomes. This is shown in figure 104.

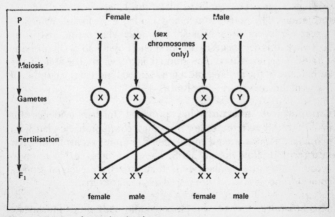

Figure 104. Sex determination in man

Evolution

The theory of evolution explains how the enormous variety of plants and animals found in the world came into existence. It explains that similarities between organisms are a result of descent from a common ancestor, and that differences between them are a result of variations accumulating between parents and offspring. It must be supposed that a structure capable of division was produced from non-living matter hundreds of millions of years ago. These relatively simple forms of life have, over millions of years, given rise by a series of small changes to a succession of living organisms which have become increasingly more varied and more complex.

Evidence for evolution
Evidence that evolution has occurred is drawn from many sources; much of it is circumstantial but it is still very convincing.

Probably the most satisfactory evidence that evolution has occurred has come from **fossils**; in addition they provide information about the succession of animals and plants over millions of years. Fossils include a variety of plant and animal remains. The most common method of fossilisation occurs when the hard parts of the body, for example shells, teeth and bones, are replaced by minerals. Whole organisms may be fossilised, for example whole mammoths have been preserved in ice in Siberia. Also imprints in rock of footprints or fern leaves are another type of fossil. There are now quite accurate methods of dating rocks and geologists have been able to place the layers in order of their formation. The most convincing evidence is found where in successive layers, organisms show gradual changes in strict time sequence; this occurs in sea urchin shells. Fossil horses also provide a regular series; there is a gradual increase in size and in the complexity of the molars, and a gradual reduction in the number of bones in the lower part of the limbs.

Comparative anatomy has provided further evidence for evolution. All vertebrates are built on a common basic plan with pentadactyl limbs, a skull and vertebral column, a central muscular heart and a hollow dorsal C.N.S. The most logical explanation to account for these similarities is that they have evolved from a common ancestor which possessed these characteristics.

The similarity of vertebrate embryos and body fluids indicates a common ancestor. The presence in the body of organs which do not function (vestigeal organs), for example the limb buds in pythons, indicates evolution from an ancestor where they did function. Further evidence is provided by the distribution of organisms and by the occurrence of living intermediates such as the duck-billed platypus which has reptilian features (lays shelled eggs) and mammalian features (mammary glands and hair).

Evolution usually takes millions of years but man can change an environment rapidly and this has provided direct evidence for evolution. Populations of insects have become resistant to D.D.T. and other insecticides. The evolution of the **peppered moth** has been most closely followed and this has provided another example of evolution in action; it is dealt with in detail at the end of the chapter.

Theory of evolution by natural selection

The man whose name is most closely associated with evolution is **Charles Darwin** (1809–82). Between 1831 and 1835 he travelled round the world on the Admiralty Research ship, H.M.S. *Beagle*, and during the voyage he collected many specimens and studied many organisms. He was particularly struck by the animals (finches, giant tortoises and iguanas) of the **Galapagos Islands** which are situated on the Equator some 900 km west of Ecuador in South America. The finches resemble those on the adjacent coast of South America, and they differed slightly from island to island. This led Darwin to believe the mainland and island species had a common ancestor but isolation on the islands caused the original stock to evolve independently as interbreeding was prevented. As a result of the voyage he came to the conclusion that animals had arisen by a process of slow and gradual change brought about by natural selection. For twenty years after his return he worked to provide evidence for his theory.

In 1858 **Darwin** and Alfred Russell **Wallace**, who independently had come to the same conclusions announced the **theory of evolution by natural selection**, and Darwin's book *Origin of Species* was published a year later. The theory of evolution by natural selection is briefly summarised.

1. Individual members of a species differ slightly from each other, that is they show variations, and some of these can be inherited.

203

2. Offspring are always more numerous than their parents.

3. Despite this tendency to increase for any one species, the numbers tend to remain more or less constant.

4. As fewer organisms live to reproduce than are produced, there must be a continuous struggle for existence.

5. Organisms which possess variations which are advantageous in a particular environment will be more fitted to live there; these will survive and reproduce and have offspring which may inherit the more successful variations. This is known as the **survival of the fittest**. There will be a tendency to diverge away from the original type by the accumulation of favourable variations.

The kind of variation which is important in evolution is that which the organism is able to hand on to its offspring and is transmitted through genes. Some of the visible variation in the phenotype results from environmental influences, particularly diet; these are non-heritable variations and they play no part in evolution. The sources of **heritable variation** were not known to Darwin; Mendel's work had not yet been published and nothing was known about genes and chromosomes. Genetic variation arises principally by mutation, by reassortment and recombination of chromosomes at meiosis, and as a result of fertilisation.

A **mutation** is a change in a gene or chromosome. Mutations are rare and most are harmful, for example vestigeal wings in *Drosophila*. Chromosome mutations usually occur during meiosis, for example when a section of chromosomes is lost. Gene mutations involve chemical changes in a gene and affect the character the gene determines; these can be caused by radiation. Variation also occurs at meiosis during the formation of gametes, when there is an exchange of material between chromosomes, and it is chance which chromatid goes to each pole. Further variation is achieved at fertilisation when genes from two parents are brought together and combined in the offspring.

The number of offspring produced by different species may vary greatly but in all cases the potential number of progeny that could be produced is enormous, for example oysters produce sixteen million eggs per year. However, population numbers for a species do not fluctuate widely, in general two individuals will give rise to two surviving offspring capable of reproducing. Large numbers of offspring must perish and there must be a continous struggle for existence. This is not always a conscious struggle and it can operate at any stage of the life history from the egg which may not be fertilised to the adult which may be infertile. In general the main factors controlling population numbers are shortage of food, oxygen, water and space, predation, disease, and in addition in plants, lack of light.

Organisms with favourable variations will stand a better chance of survival in this struggle for existence. The result is that well-adapted individuals survive, reproduce and hand on their favourable characteristics to their offspring, whereas less well-adapted individuals fail to do so. Variations occur at random and the environment acts on them.

Natural selection in action

The **peppered moth** (*Biston betularia*) exists in two forms, a normal light mottled form and a dark melanic form. The first melanic moth was reported in Manchester in 1848 but since then the melanics have virtually replaced the mottled form in industrial regions and in regions where the prevailing winds have carried atmospheric pollution. In the South-West and Ireland the mottled form remains common. The peppered moth usually rests on the branches and trunks of trees and it is predated upon by birds such as thrushes. In 1848 the trees were covered with lichens and the mottled moth was well camouflaged whereas the melanic moth was clearly visible to the predators and rarely survived long enough to reproduce. Since 1848 soot from factories has destroyed much of the lichen and darkened the surface of the trees. As a result the melanic form became almost invisible against the dark trees whereas the mottled form stood out. The dark moths are now said to have a **selective advantage** and are thus far more likely to survive, reproduce and hand on the dominant melanic gene to their offspring.

Formation of new species: isolation

A species is a group of organisms with similar features and which can interbreed to produce fertile offspring.

Organisms are not evenly distributed across their geographical range but tend to live in areas where conditions are at an optimum and thus a species becomes divided into populations. Variations will occur at random within each population, and as the conditions in which each population is living will vary there will be a different selection of variations.

As long as organisms from different populations breed and genes are exchanged the populations will all remain members of the same species. If a population becomes **isolated** and genes are not exchanged, genetic differences will accumulate and a point may be reached when the differences are so great that the members of the two populations may be unable to breed successfully. This is a stage in the formation of a new species and if isolation continues members of each population can develop into a new species.

There are many ways in which populations can be isolated. Rivers, deserts, oceans and mountains are effective barriers. Organisms may be separated by having a preference for slightly different habitats, or by breeding at different times.

Key terms

Allele One of the alternative forms in which a gene may exist.
Dominant Gene, of a pair of alleles, which is expressed in the phenotype.
Darwin, Charles 1809–82, together with Wallace, announced the theory of evolution by natural selection in 1858.
Fossils Remains of organisms preserved in rocks.
Gene A unit of genetic material which controls the synthesis of one protein.
Genetic material Handed on from generation to generation.
Genotype The genetic make-up of an organism.
Heterozygote Organism which contains two different alleles of a gene. Will not breed true for the character controlled by gene.
Homozygote Organism which contains two identical alleles of a gene. Will breed true for character controlled by that gene.
Isolation of populations; prevents exchange of genes: important in production of new species.
Mutation Spontaneous change in genetic material.
Phenotype All visible and measurable features of an organism.
Recessive Gene which in the presence of its partner allele has no influence on the phenotype.
Species Group of organisms with similar features and which can interbreed to produce fertile offspring.

Chapter 11
Ecology

This is the study of organisms in relation to their environment. A **habitat** is a specific locality with a particular set of conditions and its own **community** of organisms. These organisms are adapted to the habitat so that they are able to use the opportunities and cope with its problems. The conditions in any habitat are made up of the **physical** environment and the **biotic** environment. The physical environment includes such factors as temperature, water and light. The biotic environment consists of the interactions between organisms. A community is the name given to the collection of interacting species within a habitat.

Within a community and its environment there will be a continual flow of energy and a cycling of matter. The energy enters the system in the form of light energy which is converted by green plants into organic substances during photosynthesis. These substances are then used by all organisms in respiration and the energy released is eventually lost from the system in the form of heat. Green plants, then, may be considered as **producers** of organic matter. Animals are **consumers** since they rely on the other plants and animals for their supply of organic materials. In any habitat there are also **decomposers**, such as bacteria and fungi, which feed off the remains of plants and animals and in so doing release inorganic materials back to the soil. Through these three methods of feeding the cycle of matter is achieved; producers converting inorganic material into organic matter and decomposers finally converting organic remains into inorganic material. In any community, then, **food chains** are built up, the producers (green plants) being eaten by herbivores (plant-eating) animals, and the herbivores being eaten by carnivores (meat-eating) animals. For example, producer (grass) → primary consumer (cow) → secondary consumer (man). Food chains rarely exceed more than six links.

The number of individuals at each stage of a food chain becomes smaller, and this phenomenon is known as the pyramid of numbers. This is because the animals at the end of the chain are fewer and larger than those at the beginning and because only a small proportion of the energy taken up by each organism is handed on to the next organism in the chain. It is in fact this wastage at each stage which makes animal protein so expensive to produce, and why under-developed nations tend to rely more on crops than livestock. Food chains in this simple form rarely exist. This is because each organism usually feeds off several sources and in turn is preyed on by several organisms. This results in food webs developing in any community.

The carbon and nitrogen cycles
The constant reuse of chemical materials by living organisms is well illustrated by the carbon and nitrogen cycles. The carbon cycle is explained in figure 105.

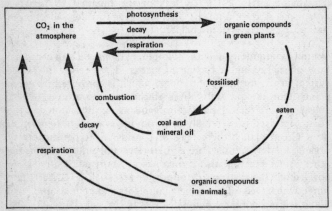

Figure 105. The carbon cycle

In the nitrogen cycle (see figure 106) de-nitrifying bacteria break down ammonium compounds in the soil resulting in the release of nitrogen into the air. This only happens in poorly aerated soils. Nitrifying bacteria may exist independently in the soil, such as *Nitrobacter* and *Nitrosomonas*, and these change the ammonium compounds into nitrates. These nitrates are then absorbed by plants. Nitrifying bacteria also exist in swellings, called nodules, on the roots of legumes (e.g. pea and clover). This is called a symbiotic relationship since both partners, plant and bacterium, benefit. These bacteria are able to use nitrogen from the air and build it into nitrogen compounds. Legumes are an important part of any crop rotation since the nodules result in them increasing the nitrogen content of the soil. Man now influences considerably the balance of nature by interfering with these cycles. For example, the amount of

208

combustion has increased considerably in the last century, resulting in the rapid depletion of fuel reserves and an increase in the carbon dioxide content of the atmosphere.

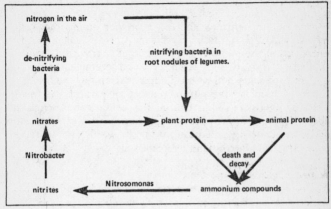

Figure 106. The nitrogen cycle

The soil

The soil acts as an anchorage for plant roots and a source of mineral salts, water and oxygen for plant growth. Soil consists of: (1) inorganic particles derived from the underlying rock by weathering. The size of the particles varies, sand being larger than clay, and this influences the amount of air and water present in a soil; (2) water, which exists as a thin film around the soil particles and is held with an increasing force the drier the soil becomes; (3) air, which occurs in the spaces between the soil particles unless the soil is waterlogged; (4) humus, the decaying remains of plants and animals which provide a supply of minerals to the soil. Humus also improves the quality of the soil by holding the particles together. (5) Soil organisms, which include bacteria, fungi and animals such as insects and worms. Bacteria are the most important of these because of their role in the breakdown of organic matter.

Heavy soils contain a high proportion of clay particles; they thus contain little air and a high proportion of water. Light soils are sandy, which means aeration and drainage are good, but there is a low water content. Minerals tend to be washed out of sandy soils more easily than clay soils. A loam is a soil which contains a

209

balanced mixture of sand and clay, and has a good humus content. It is thus ideal for plant growth.

Soil fertility may be improved or maintained in a variety of ways. **Lime** adds calcium to the soil, neutralises acid, and causes clay particles to cling together, thus improving the air content. **Manures** such as straw and faeces provide humus that yields nitrates as it rots. **Artificial fertilisers** provide elements such as nitrogen and phosphorous, but they do not improve the texture of the soil and they are easily washed out and accumulate in harmful quantities in lakes and rivers. **Crop rotation** has the beneficial effects of reducing the rate of removal of particular elements, reducing the risk of soil harbouring the parasites of one crop, and adding nitrogen to the soil when the legume crop is ploughed in and rots. **Digging** and **ploughing** kill weeds and so reduce competition for the crop, and aerate the soil.

Micro-organisms and man

Micro-organisms include bacteria, viruses and fungi (see page 235. With the exception of a small number of bacteria these micro-organisms are either saprophytes or parasites. As a result their activities are very important to man. Some of the parasitic micro-organisms cause diseases to man, his crops and his livestock, and are called pathogens (see page 235). Man has devised a number of artificial methods for controlling disease, such as immunisation, antibiotics, sewage disposal, and breeding resistant strains of plants and animals. This has resulted in increased productivity in crops and a longer life expectancy for many human beings. Unfortunately the latter also introduces fresh problems of over-population and bigger demands on the already inadequate food supply. Saprophytic organisms are important in the cycle of materials but as far as man is concerned also produce problems in terms of food spoilage. Again man has devised a wide range of techniques which can prevent this type of damage (see page 236).

Man is now the most important biotic factor in the world. He influences the environment by poisoning it and replacing the natural habitats with his own artificial environments, such as cities. This creates an imbalance which has already resulted in the extinction of many species, and which threatens the existence of man himself. Through his knowledge he can achieve good and bad effects. It is to be hoped that in the future his improved understanding of natural systems will lead to a more rational and sensitive use of his environment.

Animal and Plant Classification

Classification of animal kingdom

Protozoa Unicellular, e.g. *Amoeba, Paramecium*.

Coelenterata Body made up of two layers of cells surrounding central cavity whose only opening is the mouth, e.g. *Hydra*.

Flat worms Small flattened unsegmented bodies, e.g. tape-worm.

Segmented worms Body cavity and two openings to gut, e.g. earthworm.

Arthropoda Jointed limbs and hard exoskeleton.

    Crustacea – aquatic, many paired limbs, e.g. water-flea.

    Myriapoda – terrestrial, many segments each with similar limbs, e.g. centipedes and millipedes.

    Insecta – body in three parts, thorax bears three pairs of legs, e.g. locust.

    Arachnida – body in two parts, four pairs of legs, e.g. spiders.

Mollusca Soft bodied with one or more shells, e.g. snails, mussels.

Echinodermata Marine, radial symmetry, e.g. starfish.

Vertebrata Brain in well-developed head, skeleton of bone or cartilage of which vertebral column forms central axis. Tail.

    Fish – aquatic, move by tail and fins. Gills. e.g. shark.

    Amphibians – partially terrestrial, eggs laid in water. Moist skin, pentadactyl limb. e.g. frog, newt.

    Reptiles – terrestrial, eggs laid on land protected by shell. Scaly skins, pentadactyl limb. e.g. snakes, turtles.

    Birds – feathers, fore limb adapted for flight. 'Warm-blooded'. Eggs protected by shell. Pentadactyl limb. e.g. gulls, penguins.

    Mammals – hair, 'warm-blooded'. Young fed on milk. e.g. man, rabbit.

Classification of plant kingdom

Thallophyta No roots, stems, or leaves. Reproduce by spores.

    Algae – aquatic, contain chlorophyll, e.g. *Spirogyra*, seaweeds.

    Fungi – no chlorophyll. Saprophytic or parasitic. e.g. *Mucor*.

Bryophyta Small green plants with stems and leaves, e.g. mosses.

Pteridophyta Green plants with roots, stems and leaves. Vascular tissue. Many extinct. e.g. ferns.

Spermatophyta Reproduce by seeds.

    Gymnosperms – ovules and seeds naked, e.g. pine, larch.

    Angiosperms – ovules enclosed in ovary and seeds in fruit.

        Monocotyledons – e.g. grasses, palms, lily.

        Dicotyledons – e.g. buttercup, oak, pea.

Index

213

Examination Hints

In the biology examinations today more emphasis is placed on the understanding of basic biological principles than was the case a few years ago. With many examining boards there has been a tendency to move away from the straightforward factual questions to those of a more searching nature. Whatever the style of the questions there is no short cut to passing a biology examination; no amount of advice will help unless the syllabus has been studied, the topics understood, and revision thorough. Even this does not always produce successful examination results and the following pages indicating what the examiners look for, how the paper should be answered, and how to approach the subject, should be read carefully.

Revision

It is preferable to revise the whole syllabus. Preparing isolated topics and neglecting the remainder reduces the choice of questions to answer and leaves the possibility of passing very much to chance. It is useful to have access to a syllabus and previous examination papers; if these are studied it is possible to determine the sections of the course on which a number of questions are always set. These sections should then be revised in their entirety. If there are always a number of questions on the mammalian body, for instance, it would be pointless to revise this leaving out the heart and circulatory system. Some examining boards set a number of compulsory short questions requiring short answers. In this case it is essential to study every topic of the syllabus to ensure that they can be answered.

There is no specific method for revising; as long as the method adopted produces successful results it must be satisfactory. Students often mistake 'reading through' for 'learning'. Reading through a topic before attempting to learn it is useful to make sure it is fully understood, but few candidates would be able to answer questions in enough detail after one reading. Some students revise by writing out the work as they go through it; others read the topic through several times and then attempt to write out the main points without the aid of the books. Some students like to prepare a revision timetable based on the syllabus. As part of the revision, at the end of each topic, relevant questions can be selected and answered in the time allowed in the exam.

The written paper
Interpretation It is essential to read questions very carefully and to make sure that the meaning is fully understood before answering them. No marks are awarded for irrelevant information and the answer should include **only** points demanded by the question.

Timing It is most important to attempt the correct number of questions. If you answer only three questions instead of four, it is almost impossible to make up the marks you could have been awarded for the fourth question; it is always easier to get the first 50 per cent of the marks. It is essential to know how many questions have to be answered so that the time available for each answer can be allocated. It is most unwise to take extra time for answering a question as it is likely you will be including irrelevant material, as well as decreasing your total mark for the examination. Time should be allowed at the beginning of the examination to decide which questions are to be attempted. Where there is a choice of questions, equal marks are allocated for each answer so there is no advantage in selecting the more difficult questions.

The following paragraphs outline some of the ways in which biological questions are frequently phrased and how they should be interpreted.

Short-answer questions
These may be set as a separate examination or they may form the first part of the examination. They can take the form of one-word, one-sentence, or one-paragraph answers. A definite amount of space is left in which to answer the question, indicating the length of answer that is required. It is obvious, for example, that a one-word answer must be inadequate if four lines are given in which to answer the question.

Care must still be taken to ensure that the question is interpreted correctly. A typical short question could be, 'Give two differences in structure between arteries and veins.' Answers such as 'Arteries carry oxygenated blood, and veins carry deoxygenated blood', or 'Arteries run away from the heart and veins run towards the heart' will inevitably be given. These are differences, but they are not differences in structure. Only answers referring to the walls of arteries and veins should be given: for example, 'Artery walls contain much elastic tissue but vein walls contain very little', or 'Arteries do not possess valves; veins do possess valves'. In this type of question it is also important to refer to both the artery and the vein: for example, the answer 'Veins have valves' or 'Veins have little elastic tissue' would be unacceptable because no reference to the equivalent structure in arteries has been given.

Multiple-choice questions are part of some examination papers. Here the student is asked to tick, circle or underline the correct answer from a choice of four or five.

Example: Urea is made in:
         (a) the pancreas
         (b) the bladder
         (c) the kidney
         (d) the liver.

Do not indicate **two** right answers or the question will automatically be marked wrong. If you are not certain which is correct, choose only one answer; at least there will be a chance of getting it right and being awarded a mark.

Tabulate
A table consisting of two or more columns should be drawn up. The columns should be headed appropriately and any differing and/or similar features should be shown side by side.

Example: Tabulate the main differences between a mammal (or an animal) and a tree (or a plant).

Mammal	Tree
Heterotrophic nutrition	Autotrophic nutrition
Locomotion	No locomotion
Growth is limited and occurs all over the body	Growth is unlimited and is restricted to certain regions
Responds quickly to a brief stimulus	Responds slowly to a long stimulus
No chlorophyll	Chlorophyll
No cellulose cell walls	Cellulose cell walls
No central vacuole in cells	Permanent central vacuole in cells
Compact body	Branched body

List
Write the relevant information briefly, numbering each fact 1 ..., 2 ..., 3 ..., and so on. Questions which ask you to list the similarities and differences can be answered within a table, but it is not really correct to list or tabulate the answer if the question asks you to give an account.

Outline Only the main facts are required and elaborate detail should not be given.

Compare This should take the form of a written account which stresses the similarities and differences between the subjects given in the question.

Example: Compare the structure and function of an artery and a vein. For this question you should not write down all the characteristics of an artery followed by all the characteristics of a vein. Each point should be taken in turn and the common and differing features discussed, such as structure of the wall, the direction of blood flow, the blood pressure, the position in the body, and so on. Diagrams of sections through an artery and a vein should be drawn side by side to point out their similarities and differences.

Contrast The approach is similar to that for comparisons but the emphasis should be on their differing features.

Compare and contrast All the similarities between the subjects should be dealt with first followed by their differences.

Distinguish between This means that the differences between subjects must be shown. These questions should be dealt with in a similar way to 'contrast'-type questions.

Discuss Particular care should be taken with questions of this type. A full treatment is required but candidates often include much irrelevant information and are inclined to wander from the point. 'Discuss' frequently indicates a more critical approach and there is a possibility of expressing opinions. It is important to plan the answer, and as attempts at such questions often gain low marks they should not be attempted unless the topic is thoroughly known and understood.

Illustrated account A diagram should be supplied, followed by a written account which refers to the parts labelled and particularly to their function.

Annotated diagrams

Large, clearly-labelled diagrams should be supplied with brief appropriate notes under each label. Unless specifically asked for in the question, additional writing underneath is not required.

Diagrams

Instructions at the beginning of examination papers may tell candidates that credit will be given for labelled diagrams where relevant. In some questions diagrams will be specified, so they should be practised as part of the revision. Where not specifically required they can be included to clarify the answer or to reduce the written description: information should not be duplicated. Diagrams should be large with distinct, continuous outlines, and be clearly labelled. Labelling lines should be drawn with a pencil and ruler; they should touch the structures they are labelling, they should never cross, and they should be fairly evenly spaced around the diagram. Shading and colouring in should be avoided as they can obscure detail. Where colours or shading have been used a key should be given unless the labels have explained their significance. Blood vessels need not be drawn with double lines if red (oxygenated blood) and blue (deoxygenated blood) are used with a key. All diagrams should be at least twice the size of those in this book, which have been reduced for space reasons. Drawing many identical structures should be avoided. For example, it is a waste of time to draw every scale on a fish; only a few representative structures need be drawn very carefully.

Questions on experiments and practical work

The experiment is preferably written up under the headings – Experiment: Method: Result: Conclusion. Where appropriate a labelled diagram of the apparatus should be given and any writing on this should not duplicate the information already shown. Comments should be given where necessary of any special features of the apparatus and experimental procedure, such as how often results are recorded, how long the experiment should be run for, and so on. Most biological experiments require some form of control; a control differs from the main experiment in one respect only, that is by the one factor being investigated. Any difference in result can be attributed to the one factor that is different.

Note Longer questions on ecology require a lot of time (about twenty hours) to have been spent in a specific locality, identifying and recording the species found there. This type of work requires expert guidance and use of detailed reference books. It is beyond the scope of this book to deal fully with this topic.

Preparation of questions

It is always advisable to map out the answer before writing it in full. Frequently this procedure is neglected by candidates. With short-answer questions or where the question has been divided into parts it may not be necessary to plan the answer any further. However, in many cases, especially those questions starting with 'Discuss', 'Describe', 'Compare', or 'Contrast', the production of a brief outline can be very useful and removes the possibility of introducing irrelevancies. It also helps the candidate to present the answer in a logical sequence and lessens the chance of leaving out important points.

Before the examination, planning questions for practice can be very useful, especially in view of the fact that many questions are repeated from previous years. If papers from previous years are obtained specimen-answer plans can be mapped out.

Below is a very straightforward question and an outline answer, but without the outline important points could be omitted.

Describe the composition and functions of blood.

1. Composition

red corpuscles ⎫	structure (including shape);
white corpuscles ⎬	size; number; formation;
platelets ⎭	life span.
plasma —	constituents.

2. Function
(a) Transport of:
   oxygen   hormones
   carbon dioxide   ↑
   soluble food
   excretory products e.g. urea   heat
(b) Prevention of infection:
   white corpuscles – phagocytosis
                    – antibodies, antitoxins
   blood clotting

Practical examinations

Some examining boards set practical examinations. For these, practice is needed in observing, drawing and labelling specimens. Living material may be provided, in which case observations on behaviour will usually be required. More frequently, the questions refer only to external features of organisms, but occasionally simple dissections are required, e.g. of a flower. Where practical examinations of this nature are set, it is essential to spend some time beforehand acquiring the necessary techniques.

Specimens and questions likely to be given are:
1. Drawing and labelling external features of insects and other small invertebrates.
2. Drawing and labelling structure, both external and internal, of flowers, fruits and seeds.
3. Drawing and labelling parts of skeleton.
4. Observing and commenting on simple experiments.
5. Performing simple food tests.
6. Drawing and labelling parts of plants, e.g. bulbs, corms, rhizomes, tendrils, etc.
7. Studying the behaviour of small invertebrates, especially the movement and feeding of aquatic animals.

The same importance as for the written examination should be given to timing. It is helpful if the distribution of marks for particular questions is known so that time can be allotted accordingly.

Graphs
Many examining boards set questions on graphs and these can take two forms. Graphs may have to be constructed from data given, or questions may be asked on a graph which is provided.

To construct graphs The ability to do this is both necessary and advantageous, for many marks can be gained for technique and presentation without specialist biological knowledge. Marks can be allocated for:

full use of the graph paper

selecting the correct axis for each set of figures

dividing each axis equally into the correct number of units

labelling each axis fully and accurately

plotting each point with a small cross or a dot

joining the points

giving the graph a correct title

neatness and general presentation

Many candidates have difficulty in deciding the correct axis for each set of figures. Consider the following results of an experiment which measures the height of a plant stem in millimetres at weekly intervals.

Week	Height (mm)
0	20
1	30
2	100
3	220
4	400
5	490
6	510

The vertical axis is the y axis and the results, that is 'what the experimenter finds out' are on this axis; in this case it is the height of the stem in millimetres. The horizontal or x axis is used for units which are determined and controlled by the experimenter. In this case it is time in weeks: the experimenter decided at the beginning of the experiment how often he was going to measure the height of the stem.

Interpretation of graphs Consider the above graph. To answer questions such as 'What is the height of the stem at $3\frac{1}{2}$ weeks?', find $3\frac{1}{2}$ weeks on the x axis; with a ruler draw a line parallel to the y axis from this point upwards until it reaches the line of the graph.

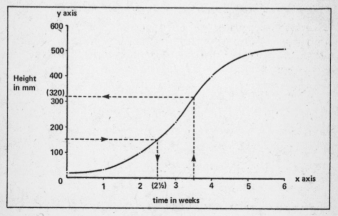

Figure 107. Correct construction of a graph

From this point draw a line parallel to the x axis across to the y axis and where this line meets the y axis is the height required; it is 320 mm. The same procedure is employed if the question asks 'At which week is the height of the stem 150 mm?'; the starting point is 150 mm on the y axis; and the answer is $2\frac{1}{2}$ weeks.

Key Facts Revision Section

Chapter 1. The cell and cell organisation

Cells are the basic units of life. They are made up of protoplasm bounded by plasma membrane. Under the light microscope protoplasm is seen as differentiated into an ovoid nucleus lying in a grey, jelly-like substance, the cytoplasm. Under the electron microscope cytoplasm is seen to be differentiated into many organelles. Within these organelles the processes of life occur. Mitochondria – site of respiration. Ribosomes – site of protein synthesis. Chloroplasts – site of photosynthesis. Endoplasmic reticulum – serves as a transport system. Centrioles – determine the position of the spindle during cell division in animals. Golgi body – secretes materials which are then transported out of the cell. Plasma membrane – the living boundary of the cell which controls the movement of substances in and out. Nucleus – contains a set or sets of chromosomes which carry the genetic material. Most organisms or cells carry two sets of chromosomes (diploid), some organisms or cells carry only one set (haploid). The number of chromosomes in a set varies depending on the species. The genetic information is in a coded form and controls the development and functioning of the cell.

Plant and animal cells are basically alike but differ in that plant cells contain a large central vacuole, chloroplasts and a cellulose cell wall. Also food reserves tend to be stored as glycogen in animal cells and as starch in plant cells. Most organisms are multicellular (many-celled) but protozoa (e.g. *Amoeba*, *Paramecium*) and some algae (e.g. *Pleurococcus*, *Chlamydomonas*) are unicellular (one-celled). In multicellular organisms cell specialisation occurs which allows for division of labour. Similar cells which carry out similar functions are grouped together to form tissues, tissues are grouped together to form organs, and organs are grouped together to form systems

Plant tissues

Vascular tissue consists of xylem vessels and phloem sieve tubes. **Xylem** vessels – rows of dead cells which have lost their cross-walls. They are lined with waterproof lignin except for small gaps in the lignin called pits. Xylem vessels transport water and mineral salts from the roots up to the leaves. They are also important in support. **Phloem** sieve tubes – rows of living cells with perforated end walls called sieve plates. These cells translocate soluble organic substances from the leaves to all other regions of the plant. **Parenchyma** – thin-walled cells with living contents. It acts as a packing tissue making up the cortex of roots, pith of stems and mesophyll of leaves. **Collenchyma** – living cells with extra bands of cellulose thickening running down the corners which give flexibility and strength to stems and leaf stalks. **Sclerenchyma** – dead elongated fibres. The cells are thickened with lignin and give strength to stems. The **epidermis** – a protective layer over young regions of the plant. The cells fit closely together and are often covered by a waterproof cuticle. Gaps or stomata occur in the epidermis to allow gas exchange. In older regions the epidermis is replaced by a waterproof bark, perforated by lenticels for gas exchange.

Chemically, cells consist of organic and inorganic compounds. Organic compounds are complex molecules containing carbon and include carbohydrates, fats, proteins, vitamins and nucleic acids. Inorganic compounds are simple and include water and mineral salts. **Carbohydrates** – these include sugars and starch. Their function is to supply a store of energy for the organism. They are also a structural component in some cells – e.g. cellulose wall of plant cells. The simplest carbohydrate is glucose, $C_6H_{12}O_6$ and is the building block for more complex carbohydrates. Starch and glycogen are stored since they are insoluble and do not interfere with the osmotic pressure of the cell. **Fats** are stored as an energy reserve and insulating layer. They are insoluble and made up of fatty acids and glycerol. **Proteins** contain the elements carbon, hydrogen, oxygen and nitrogen, and sometimes phosphorus and sulphur. They are built up from about twenty amino-acids. They are used for growth and replacement of cells. Also all enzymes are proteins. Enzymes are organic catalysts, speeding up chemical reactions in living cells. Each reaction has a specific enzyme which will only function within a narrow range of temperature and pH. **Vitamins** are chemically complex but only needed in minute quantities. They are involved in a variety of functions (see vitamins, page 227). **Nucleic acids** include deoxyribonucleic acid (DNA) and ribonucleic acid (RNA). DNA only occurs in the nucleus, being combined with protein to form chromosomes. DNA carries the genetic information in a coded form. The unit of the code is a sequence of three bases, chosen from four possible bases. These varied triplets run down the length of the DNA molecule and produce an almost infinite number of combinations. RNA acts as a messenger, carrying the coded information from the nucleus to the cytoplasm. The code is translated into the synthesis of specific enzymes, and this occurs at the ribosomes. **Water** makes up 60–90 per cent of most cells. It is vital for life because as well as being the major constituent of protoplasm it acts as a solvent, substances being transported and excreted in solution. It cools the body, is a reactant in photosynthesis and is the medium in which all chemical reactions occur. **Mineral salts** are chemically simple and needed only in small quantities by organisms. They are used for a variety of functions (see page 227).

Mitosis

This is a form of cell division in which the chromosome number is kept constant. It occurs in growth and the genetic material is copied exactly. It is a continuous process but is divided into several phases. **Interphase** (resting phase) – the chromosomes are not visible but are carrying out normal cell activities. **Prophase** – the chromosomes, consisting of two identical chromatids joined at the centromere, become visible due to coiling; spindle threads begin to form and the nuclear membrane breaks down. **Metaphase** – the chromatids move to the equator of the cell and become attached to the spindle. **Anaphase** – the chromatids separate from each other and move to opposite ends (poles) of the cell. **Telophase** – the two groups of chromatids reform as chromosomes, they become invisible and a nuclear membrane forms around them. The cytoplasm divides by constriction in animal cells, but in plant cells a new cell wall develops between the two nuclei.

Meiosis

During a life cycle involving sexual reproduction the process of cell division called **meiosis** occurs at some stage. As a result of this process four daughter cells are produced which contain **half** the number of chromosomes of the parent cell and are not genetically identical. Meiosis is also divided into a number of stages, although it is a continuous process. Meiosis I involves the pairing of homologous chromosomes to form a unit called a bivalent. During this stage crossing over of chromatids occurs between the chromosomes forming an arrangement called a chiasma. Homologous chromosomes position themselves on the equator of the spindle and then separate, moving to the poles. During Meiosis II chromosomes move to the equator and the chromatids then separate to the poles.

Transport

Within a cell, transport may be by the physical processes of diffusion and osmosis, or by active transport. **Diffusion** – atoms and molecules are in constant motion as an effect of heat, and this means that when they are in a gas or liquid form they are able to move about. Thus they may become evenly distributed through a space, moving from a high concentration to a low one. **Osmosis** – a special case of diffusion when, because of a semi-permeable membrane, water molecules only can move. In a dilute solution there are more water molecules than in a concentrated solution, so water diffuses from the weaker to the stronger. **Active transport** involves the expenditure of energy by the cell to move substances in or out of the cell, often against the diffusion gradient.

Multicellular plants rely on osmosis to move water relatively short distances, e.g. across the root and leaf. The ability of a plant to take in water is called suction pressure (S.P.). S.P. is a balance between the force tending to push water in, osmotic pressure (O.P.) and the force tending to prevent water from entering, the wall pressure (W.P.). S.P. = O.P. – W.P. A cell with no ability to absorb water (S.P. = O) is called fully turgid.

Chapter 2. Nutrition

This is the method by which organisms obtain or produce organic substances; the method of nutrition is a basic difference between animals and plants.

Autotrophic – involves the synthesis of organic compounds from inorganic materials. Usually light energy is used to build up CO_2 and H_2O into complex foods. This is the characteristic method of feeding in all green plants (holophytic) and involves the process of photosynthesis.

Heterotrophic – the method of feeding of all organisms unable to synthesise their own food and so rely on the environment producing organic substances. Characteristic of all animals and fungi and most bacteria. Subdivided into **holozoic** – feeding on solid organic materials; **parasitic** – feeding off organic compounds present in the body of another organism; **saprophytic** – feeding off soluble organic compounds from remains of plant and animal bodies.

Photosynthesis – represented by the equation $6CO_2 + 6H_2O \xrightarrow[\text{chlorophyll}]{\text{light}} C_6H_{12}O_6 + 6O_2$ Light energy is trapped by chlorophyll and used to split water into hydrogen and oxygen. The oxygen is given off as a waste product and the hydrogen is added to carbon dioxide from the atmosphere to produce sugars, complex carbohydrates and fats. Proteins are finally synthesised using other elements such as nitrogen and sulphur taken up from the mineral salts in the soil.

Structure of the leaf Chlorophyll occurs mainly in leaves, contained inside chloroplasts; the structure of leaves allows photosynthesis to occur efficiently.

Factors affecting photosynthesis Chlorophyll, CO_2, light and water are all needed. This is shown (except for water) in simple experiments depriving leaves of **one** factor. Role of water can only be shown using radioactive isotopes. Rate of photosynthesis is shown by measuring release of oxygen bubbles from water plants. Essential that photosynthesis occurs more rapidly than respiration if growth is to occur: when they proceed at same rate called compensation point.

Heterotrophic nutrition
A balanced diet for man (omnivore) will include the following. **Carbohydrates** Contain carbon, hydrogen, oxygen, with hydrogen and oxygen in the proportion 2:1. They are used in respiration and form food stores. Three main groups: **monosaccharides** – sweet, simple sugars, e.g. glucose, fructose, $C_6H_{12}O_6$; **disaccharides** – sweet, double sugars, e.g. sucrose, maltose, $C_{12}H_{22}O_{11}$; **polysaccharides** – many sugar molecules joined; insoluble and used for storage, e.g. starch, glycogen. Carbohydrates are provided by flour, sugar, potatoes.

Proteins Contain carbon, hydrogen, oxygen, nitrogen and often sulphur and phosphorus; these are linked to form **amino-acids** of which there are about twenty-two known types. Most proteins contain hundreds of amino-acids and different arrangements of them form different proteins. Proteins are used for cell, enzyme and hormone synthesis and indirectly as a source of energy. Fish, meat and eggs are rich in essential amino-acids.

Fats Contain carbon, hydrogen, oxygen (less than carbohydrates). Used in respiration and as food stores as they contain twice as much energy per gramme as carbohydrates.

Vitamins Complex chemicals required in small amounts for essential reactions. Gowland Hopkins realised importance.

	Source	Function	Effect of lack
A	Fish-liver oils, liver, milk, carrots (carotene)	Skin growth, night-vision	Dry skin and cornea, night blindness
B.B$_1$ B$_6$	Thiamine; yeast, wheat germ Nicotinic acid; yeast, meat	Helps energy release Helps energy release	Beri-beri; nerve, muscle wasting Pellagra (skin, gut nerve disorders)
C	Oranges, lemons, blackcurrants	Skin growth, capillary strength	Scurvy (skin sores, poor healing, capillary bleeding)
D	Fish-liver oil, egg yolk, action of sun on skin	For use of calcium and phosphorus	Rickets; poor bone development

Mineral salts Simple chemicals required in varying amounts for chemical activities and tissue construction. **Sodium** and **chlorine** taken in as sodium chloride (cooking salt) needed for body fluids. **Sodium** and **potassium** from plant food, for nerve and muscle action. **Calcium** and **phosphorus** from milk and meat, for bones and teeth. **Iodine** from water and fish, required for the thyroid gland; lack causes 'simple' goitre. **Iron**, from liver, for haemoglobin; lack causes anaemia.

Water Acts as a solvent, forms body fluids and protoplasm, important in enzyme action.

Roughage Cellulose from plant cell walls; stimulates peristalsis.

Food tests

Food	Test	Positive result
Carbohydrates 1. Reducing sugar (glucose)	Add Benedict's solution; boil.	Brick red precipitate. Green (little sugar)
2. Non-reducing sugar (sucrose)	If 1. is −ve, add dil. HC1, boil for 2 mins. Add solid sodium bicarbonate until fizzing stops. Repeat 1.	Brick red precipitate. Green (little sugar)
3. Starch	Iodine solution.	Blue/black
Protein	Add Millon's reagent; boil.	Pink/red
Fat	Shake food with 2ml alcohol; pour off alcohol into 2ml water. Shake.	Permanent cloudiness

Digestion and absorption of food

Digestion is the breaking down of large complex molecules into soluble molecules so they can be absorbed into the blood and used in the body. It is chemical (enzymes) and mechanical (teeth and muscles). Man is an omnivore.

Regions of the alimentary canal

Mouth Incisor teeth bite off the food and it is chewed by the premolars and molars and mixed with saliva which lubricates and moistens the food, and contains the enzyme **ptyalin** which converts starch to maltose. The tongue forms the food into a bolus and this is pushed to the back of the mouth and swallowed. It passes down the oesophagus by **peristalsis** (circular muscles contract behind the food) and into the stomach.

Stomach Food is retained here for two to four hours. Stomach muscles churn up the food with gastric juice from the stomach wall to form chyme. The juice contains the enzyme **pepsin** which converts protein to peptones, and hydrochloric acid which activates pepsin, provides the correct pH, and kills bacteria. In young children the enzyme **rennin** clots milk. The chyme is released into the duodenum (first part of small intestine) at intervals. **Duodenum** receives **bile** from the liver (stored in the gall bladder) via the bile duct, and pancreatic juice from the pancreas. Bile neutralises acid chyme, emulsifies fat and provides the correct pH. Pancreatic juice contains three enzymes: **trypsin**, converts proteins to peptides; **lipase**, converts fats to fatty acids and glycerol; **amylase**, converts starch to maltose. Enzymes from the intestine wall complete digestion: **erepsin** converts peptones to amino acids, **lipase** (as above), **sucrase** converts sucrose to glucose and fructose, and **maltase** converts maltose to glucose. Lactase converts lactose to glucose and galactose. Absorption occurs.

Colon or large intestine absorbs water from undigested food. Bacteria produce vitamins. **Rectum and anus** Rectum stores faeces (undigested food and bacteria) before their expulsion through the anus.

Absorption of soluble products This occurs through the wall of the **ileum** which has villi to increase its surface area. Amino acids, glucose, and fatty acids and glycerol pass into the blood capillaries (energy is required) and are taken to the liver in the hepatic portal vein. Fat droplets pass into lacteals which are branches of the lymphatic system.

Functions of the liver

1. Regulation of blood sugar level.
2. Deamination of excess amino-acids, with formation of urea.
3. Breakdown of red blood corpuscles.
4. Storage of iron.
5. Storage of fat-soluble vitamins, e.g. A and D.
6. Formation of bile.
7. Production of heat which is distributed to rest of body.

8. Production of plasma proteins, e.g. fibrinogen (for clotting).
9. Regulation of fats in the body.
10. Detoxication of poisonous compounds.

Chapter 3. Respiration: gaseous exchange

Occurs in every living cell and is the method by which energy is made available for processes such as movement and protein synthesis. Represented by $C_6H_{12}O_6 + 6O_2 \rightarrow 6H_2O + 6CO_2 + 2880$ kJ. Respiration can be considered in two stages: (1) **External respiration** involves gas exchange and transport of gases between the cells and respiratory surface. It is highly variable between organisms; (2) **Internal respiration** – main features are common to plants and animals. Occurs in two stages, the first not requiring oxygen (anaerobic) and the second requiring oxygen (aerobic). In the first stage the glucose molecule (6C) is broken down into two molecules of pyruvic acid (3C) and a small amount of energy is released. In the second stage the pyruvic acid is completely broken down into carbon dioxide and water and a larger amount of energy is released. The energy released in both stages is used to build up the energy-rich molecule adenosine tri-phosphate (ATP). ATP breaks down easily and makes energy available for any process which requires it in the cell. Some organisms can exist without oxygen for long periods of time. Yeast for e.g. converts pyruvic acid into alcohol and carbon dioxide. The process is called fermentation and is represented by $C_6H_{12}O_6 \rightarrow 2C_2H_5OH + 2CO_2 + 210$ kJ. Some organisms, e.g. mammals, can survive short periods of time relying only on anaerobic respiration. This results in lactic acid accumulating in muscles during periods of strenuous exercise.

Gaseous exchange in animals

This occurs by **diffusion** (movement of molecules of a liquid or gas from a region of high concentration to a region of low concentration) across a **respiratory surface**; these are thin, moist, permeable, have a large surface area, and are usually well ventilated and have a good blood supply. In *Amoeba* the surface area to volume ratio is large enough for diffusion across the cell membrane to be adequate, and this forms the respiratory surface. In larger, more active organisms a specialised respiratory surface is developed in a certain region, and there are several types, e.g. the skin (epidermis) in earthworms, the tracheal system with tracheoles in insects, gills in fish, the skin, buccal cavity and lungs in frogs, lungs and air sacs in birds, and lungs in mammals.

Mammals Air is normally drawn in through the nostrils, is warmed, moistened and filtered in the nasal cavity, and passes through the pharynx, glottis, larynx, trachea, bronchi, bronchioles and into the air sacs or alveoli of the lungs. To **inhale**, the pressure in the thorax must be reduced. The **diaphragm** muscles contract and it flattens, the **intercostal** muscles contract and the rib cage moves up and out; this increases the volume of the thorax, the pressure is thus reduced, the lungs inflate and air is drawn in. To **exhale** the reverse of this process occurs.

Alveoli have a large surface area, and they are thin, moist and covered with blood capillaries. Oxygen dissolves in the fluid lining the alveoli, diffuses into the blood capillaries, combines with **haemoglobin** and is transported as oxyhaemoglobin to the cells. This dissociates and oxygen diffuses into the cells, internal respiration occurs and carbon dioxide diffuses out. This is carried to the lungs mainly as **bicarbonate ions** dissolved in the plasma; carbon dioxide is released in the lung capillaries, diffuses into the alveoli and is expired.

Chapter 4. Transport

The blood of man consists of the following.
Red corpuscles (erythrocytes) These are minute bi-concave discs which contain **haemoglobin** but no nucleus. They are made in the bone marrow and after 100 to 120 days are destroyed in the liver or spleen. **White corpuscles** (leucocytes) There are 600 red: 1 white cell. **Polymorphs** have an irregular nucleus, a changeable shape and are made in bone marrow. **Lymphocytes** have a regular nucleus and are made in the lymph glands. **Platelets** Small fragments of cells produced in bone marrow and concerned in clotting. **Plasma** A yellow alkaline fluid, consisting of 90 per cent water and in which soluble food, fibrinogen, hormones, antibodies and antitoxins are dissolved.

Function To maintain a constant internal environment:
by transporting substances
1. Oxygen from lungs to tissues as oxyhaemoglobin.
2. Carbon dioxide from tissues to lungs in plasma.
3. Excretory products from tissues to kidneys, etc.
4. Hormones from endocrine glands to tissues.
5. Heat from tissues to the rest of the body, e.g. skin.
6. Soluble food from gut to liver to tissues.
by preventing infection. Clotting seals wounds against the entry of micro-organisms.
Antibodies act against bacteria. See page 239.

Circulation of blood is maintained by the **heart**, a double muscular pump, separated into right and left halves and carrying deoxygenated and oxygenated blood respectively. Each half consists of an upper, thin-walled **auricle**, which receives blood, and a lower, thick-walled **ventricle** which distributes it. Valves between the auricles and ventricles (**tricuspid** and **bicuspid**), and at the end of the ventricles (**semilunar**), prevent the backflow of blood.

Arteries have thick, muscular, elastic walls, a narrow bore, no valves, and carry blood with a high, fluctuating pressure away from the heart. **Veins** have thinner walls, a wide bore, valves and carry blood with a low, even pressure back to the heart. **Capillaries** connect arteries and veins and have very thin walls. Plasma passes through them, bathes the cells and allows exchange of substances between the blood and the tissues. Excess fluid drains into lymph vessels and becomes **lymph** which returns to the blood system at the subclavian veins in the neck.

Transport in plants
Entry of water Root hairs present a large surface area for absorption. Water enters them by **osmosis**; this is the movement of water molecules from a dilute solution to a more concentrated solution through a semi-permeable membrane. The soil water forms a dilute solution, the cytoplasmic lining of the root hair is semi-permeable, and the cell sap of the vacuole forms a concentrated solution. Thus water enters the root hair cell by osmosis; its cell sap is now more dilute than the sap of the next cell (of the cortex), so water leaves the root hair cells by osmosis and passes into the next cell, making that more dilute. In this way water passes across to the xylem and into it (not by osmosis) as the contents of the xylem are flowing upwards. Xylem consists of long columns of dead cells impregnated with lignin; it is important in support.

Transpiration and passage of water Transpiration is the evaporation of water from the aerial part of the plant and occurs mainly via the stomata of the leaves. The mesophyll cells in the leaf are in contact with air spaces, so water evaporates from them and the water vapour diffuses out through the stomata. These cells are now more concentrated so they take water from their neighbours which in turn draw water from the xylem. This causes water to be pulled up the xylem. This is the transpiration pull, and is helped by root pressure and osmotic push from the root cells. Transpiration rate depends on the number and position of the stomata and on atmospheric conditions (temperature, humidity, wind). It is rapid in dry, warm, moving air in the light. It is measured using a potometer.

Uptake of mineral salts Root hairs absorb ions of salts from soil water, they are then transported in the transpiration stream. As the concentration of many salts in root hairs is often greater than in soil water diffusion cannot occur. As ion uptake is faster when oxygen is present it is thought that it is an active, energy-requiring process. Nitrogen, phosphorus and sulphur (absorbed as nitrates, phosphates and sulphates) are required for protein synthesis; iron and magnesium are required for chlorophyll; calcium is required for cell walls. They are all carried in the xylem.

Translocation Transport of organic substances is carried out by phloem cells which are living and carry this soluble food actively from stores or leaves to growing or respiring regions. The mechanism of the process is not fully understood.

Chapter 5. Excretion, osmo-regulation and temperature regulation
Excretion is the removal from the body of the waste products of metabolism, so keeping the internal composition of the body constant.

Excretory products are: water and **carbon dioxide** from respiration; nitrogenous compounds, e.g. **urea**, from breakdown (deamination) of excess amino-acids in the liver; excess **salts** and **toxic** substances. **Excretory organs** are as follows.

1. **Lungs** Excrete carbon dioxide and water vapour.
2. **Kidneys** By excreting urine containing water, salts, urea and toxic substances, they act as **osmo-regulatory organs**. Osmo-regulation is the process by which the osmotic pressure of the blood and body fluids is kept constant. The kidneys are paired and contain millions of filtering tubules (**nephrons**). Blood containing waste products enters the kidney via the **renal artery** and passes into **glomeruli** (knots of blood capillaries), and blood pressure 'forces' small molecules, e.g. water, glucose, urea, into cup-shaped **Bowman's capsules**. This process (**pressure filtration**) does not require energy. **Selective reabsorption** occurs as the filtrate passes down the tubules and **loops of Henle**; all soluble food, most of the water and necessary salts are reabsorbed back into the blood. This process requires energy. The **urine** now passes down the **ureter** and is stored in the **bladder**. The blood, now with its correct composition, enters the **renal vein**.

3. **Skin** The upper **epidermis** consists of an impermeable cornified layer, a granular layer, and a Malpighian layer of actively dividing cells which gives rise to hair and sweat glands. The lower **dermis** contains receptors, nerves, and blood capillaries. The **sweat glands** act as excretory organs by extracting water, salts and some urea and passing them to the surface as sweat.

The **skin** is important in **temperature regulation**. Heat is produced in active tissues, e.g. liver, muscles, and taken to the skin by blood. To increase heat loss **vasodilation** occurs; capillaries dilate so more blood passes to the surface and more heat is lost by radiation. Sweat production increases and latent heat for its evaporation is taken from the body. To reduce heat loss, **vasoconstriction** occurs; capillaries narrow so less blood flows to the surface. Sweat production decreases. Hair is erected trapping a thicker layer of air (a bad conductor), and heat loss is reduced.

Chapter 6. Locomotion and support

Animals move to find food by muscles contracting against a skeleton. Three types: **hydrostatic skeleton** (earthworms) – fluid in the body under pressure; **exoskeleton** (insects) – hard material outside the body; **endoskeleton** (vertebrates) – hard material within the body. In man the endoskeleton is made of bone and a little cartilage held together by ligaments. It has two main regions: axial skeleton (skull, vertebral column, rib cage); appendicular skeleton (girdles and limbs).

Axial. Skull consists of the cranium to protect the brain, capsules to protect the sense organs, jaws with teeth. The **vertebral column** (man) consists of 33 vertebrae; typically each has a ventral centrum, a neural arch around the spinal canal, a neural spine, transverse processes and zygapophyses. Grouped by structural differences for specific functions. **Cervical** (7): support the neck, have a vertebrarterial canal and the **atlas** (first), and **axis** (second), carry the head and allow it to nod and rotate. **Thoracic** (12): have facets for rib articulation; large neural spines for attachment of back and shoulder muscles. **Lumbar** (5): heavy and weight-bearing with large zygapophyses for muscles. **Sacral** (5 fused): attached to the pelvic girdle and transmits movement from legs to body. **Caudal** (4 fused): reduced and functionless. **Appendicular**. Pectoral girdle consists of the **clavicle** and the **scapula** which articulate with the forelimb. Pelvic girdle consists of two innominate bones which are fused to the sacrum, and articulate with the hindlimbs. Limbs based on a **pentadactyl** (5-fingered) plan.

Functions of the skeleton
1. **Support**: maintenance of body shape.
2. **Protection**: skull protects the brain and sense organs, ribs protect heart and lungs, vertebrae protect spinal cord.
3. **Locomotion. Joints**: formed where bones meet. In **immovable** joints bones interlock for

extra strength, e.g. in the skull. **Movable** joints are **synovial**; they are enclosed in a capsular ligament and lined by a synovial membrane which secretes lubricating synovial fluid. Types are **ball and socket**, e.g. hip, shoulder; allows rotational movement (up to 360°): **hinge**, e.g. knee, elbow; movement in one plane only (up to 180°). **Muscles**: attached to bones by **tendons**. They work in **antagonistic pairs** by **pulling** on bones, e.g. the **biceps** muscle of the arm contracts, the **triceps** muscle relaxes, and the elbow is bent; the triceps contracts, the biceps relaxes and the elbow is straightened. Muscles change their shape but not their volume; they act across joints as **levers**.

Plant support The stem is an organ of elongation and requires support. Woody stems rely entirely on supporting tissues containing xylem vessels and fibres, both of which have lignified walls. Herbaceous stems have some lignified tissue but also rely on the turgor pressure of the cells.

Chapter 7. Co-ordination

An organism's metabolism consists of hundreds of chemical reactions which must be controlled and linked if the organism is to function efficiently. Also each organism must be able to take advantage of the favourable aspects of the environment and, in the case of animals in particular, move away from unfavourable aspects. Co-ordination and sensitivity are characteristics of all organisms and are carried out by means of the nervous and endocrine systems in animals and hormones in plants.

Co-ordination in plants
The many aspects of plant growth and development are controlled entirely by hormones. This means that generally plants respond slowly to prolonged stimuli in the environment.

Tropisms are permanent growth responses of a plant organ, the direction of which is controlled by the direction of the stimulus. Response may be positive (towards the direction of the stimulus) or negative (away from the direction of the stimulus). Stimuli may be light (phototropism), gravity (geotropism), water (hydrotropism). Tip of organ detects stimulus and secretes hormone which diffuses back to region of elongation where response occurs. Higher levels of hormone cause growth to be stimulated in stems and inhibited in roots. Hormone involved named auxin by Went; since chemically identified as indole acetic acid (IAA).

Co-ordination in animals
Endocrine system   Hormones are chemicals secreted from endocrine glands into the blood. They affect parts or the whole of the body. **Pituitary gland** (at the base of the brain), produces many hormones. Some control the activities of other glands, others affect growth rate and reproduction. **Thyroid gland** (on the larynx) produces **thyroxine** which controls growth and oxidation of glucose. **Adrenal glands** (above the kidneys) produce **adrenaline** in times of stress. It releases glucose into the blood, increases heart and breathing rate, sends blood to muscles. **Pancreas** produces **insulin** which controls the conversion of glucose to glycogen. Failure to do so causes excess glucose in the blood and the disease diabetes results.

Nervous system   Consists of the central nervous system (brain and spinal cord), and the peripheral nervous system (cranial and spinal nerves). The units of the nervous system are **neurones**; each consists of a cell body and two or more fibres; dendrons carrying impulses to the cell body, axons carrying them away. Adjacent fibres communicate across **synapses**. The **brain** receives impulses from the anterior sense organs and from other parts of the body via the spinal cord, and selects the best response. The main regions are the cerebral hemispheres, cerebellum and medulla oblongata. The **spinal cord** consists of neurones (grey matter) and nerve fibres (white matter), and from it paired spinal nerves arise.

A **reflex action** is a rapid automatic response to a stimulus (a change in the environment). Structures involved in a reflex arc are **receptors** (which detect the stimulus), **sensory neurones** (carry impulses to the C.N.S.), **intermediate neurones** (in the C.N.S.), **motor neurones** (carry impulses from the C.N.S.), **effector organs** (muscle or gland). In a **conditioned reflex** an irrelevant stimulus replaces the normal one.

The eye
The anterior chamber is covered by a transparent **cornea** and contains **aqueous humour**, a muscular **iris** and its aperture the **pupil**, and a biconvex **lens** attached to **ciliary bodies** by

suspensory ligaments. The posterior chamber has an outer **sclerotic coat**, contains **vitreous humour**, and is lined by a vascular **choroid** and light sensitive **retina**. This contains **rods** and **cones**; cones are sensitive to colour and are concentrated at the **fovea**, the region of most distinct vision. Nerve fibres leave at the **blind spot** as the **optic nerve**.

Vision Light rays are refracted by the cornea, humours and lens, and are focused on the retina. The rods and cones are stimulated and impulses pass to the brain giving the sensation of vision. To form clear images of near and distant objects the focal length of the lens is altered (**accommodation**). At rest the eye is focused for distant objects. To focus near objects, the ciliary muscles contract, the pull on the suspensory ligaments is released, and the lens becomes thicker. The iris controls the amount of light entering the pupil.

The ear

The outer ear consists of a **pinna**, an **auditory canal** and an **ear drum**. The air-filled middle ear contains the **ear bones** (malleus, incus and stapes). The fluid-filled inner ear contains the **cochlea, semi-circular canals, utriculus** and **sacculus**. From them branches of the **auditory nerve** pass to the brain.

Hearing The pinna directs sound waves on to the ear drum which vibrates. Vibrations are magnified by the ear bones and transmitted to the oval window which sets up vibrations in the perilymph. This causes fibres of the appropriate wave length in the **cochlea** to resonate, sense cells are stimulated and impulses are passed to the brain and interpreted as sound.

Balance Gravity is detected by **utriculus** and **sacculus** which contain endolymph and chalk grains attached to sensory hairs. Tilting causes the grains to pull on the hairs and impulses to the brain give awareness of position. **Head movement** is detected by **semi-circular canals** which contain endolymph and sensory hairs at their base. Movement causes pressure on different sensory hairs, and impulses to the brain give the sensation of movement.

Chapter 8. Growth and development

Growth is a permanent increase in size and is often linked with development. It can be measured in a number of ways. Animal growth takes place throughout the body but plant growth is restricted to regions called meristems. Many animals stop growth at maturity but most plants continue to increase in size throughout their life. Growth is influenced by the internal factors of genes and hormones and the external factors such as light, temperature and nutrition.

Insects grow by a series of moults which involves shedding the existing cuticle and the body increasing in volume as the new cuticle expands and hardens. **Metamorphosis** is a change in form during development, involving the method by which a larva becomes an adult. In some insects complete metamorphosis occurs in the life cycle (egg→larva→pupa→adult); during the pupa stage the larva becomes a strikingly different adult. Incomplete metamorphosis occurs when eggs develop into adults by a series of nymphs, which differ only slightly from the adult. In amphibia metamorphosis is associated with the gradual adaption of the aquatic tadpole into the semi-terrestrial adult.

Chapter 9. Reproduction

Asexual reproduction does not involve the fusion of gametes. It is common in plants and lower animals, e.g. *Amoeba*. The offspring are genetically identical to their parents so variation does not occur, but it allows rapid colonisation of new areas. In flowering plants it is called vegetative propagation; new plants form by separation of part of the vegetative structure, e.g. bulbs, corms.

Sexual reproduction involves the fusion of **gametes**; this is **fertilisation** and the resulting **zygote** develops into a new individual. It is common in plants and animals, and produces variation which is important in evolution. Male gametes are **sperm** (numerous, motile, small). Female gametes are **ova** or eggs (few, non-motile, large as they contain food).

Sexual reproduction in man

Female – The eggs are produced in **ovaries** and released singly into the **oviduct** at monthly intervals. At this time the **uterus** or womb is preparing to receive a zygote; if one does not arrive the lining is shed (**menstruation**). **Male** – Sperm are produced in **testes** which lie in scrotal sacs. During intercourse the erect **penis** is inserted into the vagina and sperm pass along the **vas deferens**, receive secretions from glands to form **semen**, pass down the urethra, and are deposited at the top of the vagina. They swim through the uterus and up the oviduct, where if an egg is present fertilisation will occur. The zygote passes down the oviduct and becomes implanted in the uterus wall where the **placenta** develops. The embryo is linked to the placenta by the **umbilical cord**. The embryonic blood passes to and from the placenta and here the maternal and embryonic blood come into close contact (but never mix). Dissolved food and oxygen pass to the embryo, and wastes and carbon dioxide pass to the mother. A **water sac** surrounds and protects the embryo. After nine months (**gestation**), birth occurs.

Sexual reproduction in plants

Spirogyra – filamentous alga consisting of long unbranched strands made up of identical calls. Common in stagnant pools where it floats in bright green masses on top of the water. Sexual reproduction occurs between closely packed strands and is called conjugation (see figure 108). The cell contents which move through the conjugation tube are called male gametes and the stationary cell contents are called female gametes. The resulting zygospore is resistant to desiccation and cold, and in favourable conditions will germinate to produce vegetative cells.

Mucor – saprophytic fungus which commonly grows on bread, jams, etc. Mass of vegetative hyphae, called the mycelium, has a white fluffy appearance. Sexual reproduction usually occurs when conditions are becoming unfavourable. (e.g. food being used up). Branches form between hyphae in contact with each other. The tips of the branches swell and become separated off by the formation of cross-walls. The protoplasm from the two tips mixes and a resistant zygospore is formed (see figure 108). This will germinate in favourable conditions to form immediately an erect hypha with a spore case on its end in which asexual reproduction occurs.

Angiosperms – flowering plants which are the most highly evolved terrestrial plants. The reproductive organs are contained within the flower. Each flower consists of four whorls of modified leaves arranged on a receptacle.

Calyx of sepals, the outer whorl protecting the bud.

Corolla of petals, which attracts insects.

Androecium of stamens, each with a filament and an anther producing pollen grains (contain male gametes).

Gynaecium of carpels, each with a stigma, style and ovary in which the ovule contains the female gamete.

Pollination – the transfer of pollen from the anther to the stigma of the same flower (self), or to another flower of the same type (cross).

Insect-pollinated flowers, e.g. buttercup, have scented, brightly coloured corollas often with nectaries. Pollen grains are large and rough, and stigmas are sticky.

Wind-pollinated flowers, e.g. grasses, are inconspicuous with petals green or absent. Anthers (feathery) and stigmas hang outside the flower. Pollen grains are light, small, smooth, dry and numerous.

Fertilisation Pollen grains germinate on the sugary stigma and a pollen tube grows down the style, towards the ovary and into the ovule. Male nucleus passes down the tube and fuses with the egg cell to form a zygote. The fertilised ovule becomes a seed, and the whole ovary becomes the fruit.

Pollination mechanisms Flowers have a variety of methods to ensure cross-pollination (see figure 109). **Buttercup** – the position of the anthers changes as the flower ages, falling outwards towards the petals. This means that in young flowers insects touch the anthers and collect pollen on their backs but in older flowers touch the stigma. In the **white deadnettle** although the stigmas and anthers ripen at the same time one of the branches of the stigma curves down below the anthers. This means that it is touched first by the bee and must receive any pollen it is already carrying. The **plantain** is wind-pollinated but the stigma ripens before the anthers for cross-pollination.

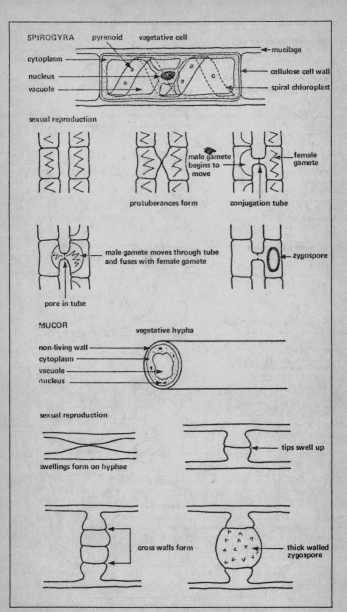

Figure 108. *Sexual reproduction in* Spirogyra *and* Mucor

235

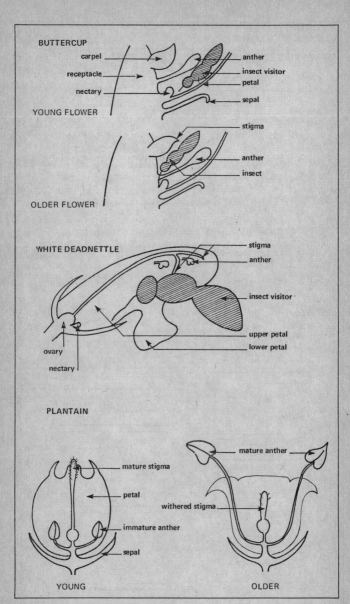

Figure 109. Pollination mechanisms

236

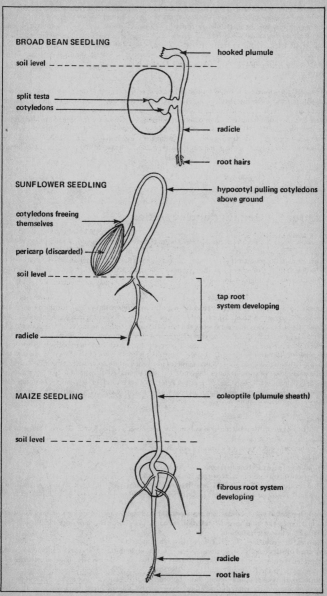

BROAD BEAN SEEDLING

soil level — — — — — — — — — — — — — — —

hooked plumule

split testa

cotyledons

radicle

root hairs

SUNFLOWER SEEDLING

hypocotyl pulling cotyledons above ground

cotyledons freeing themselves

pericarp (discarded)

soil level — — — — — — — — —

tap root system developing

radicle

MAIZE SEEDLING

coleoptile (plumule sheath)

soil level — — — — — — — — — — — — — — —

fibrous root system developing

radicle

root hairs

Figure 110. Types of germination

237

Fruit and seed dispersal Wide dispersal reduces competition, prevents overcrowding and promotes colonisation of new areas. **Wind dispersal** – light seeds or fruits with a large surface area so the wind can carry them. Dandelion fruits have a parachute of hairs. Sycamore fruits have a 'wing'. **Animal dispersal** – hooked fruits (goosegrass) may be carried externally on fur or feathers and drop off later. Succulent fruits (strawberry) may be swallowed and the seeds passed out unharmed in the faeces, many miles away. Thrushes disperse mistletoe, squirrels disperse acorns. **Dispersal by explosion** – pods of lupin and gorse split suddenly when dry and the seeds are thrown out.

Seeds and germination A seed is protected by a coat or testa and contains an embryo and a food store which may be in the cotyledons. For seeds to germinate, water, oxygen and a suitable temperature are required. Water intake causes the seed to swell and it activates the enzymes which convert the insoluble food reserve into soluble substances. These are transported to the radicle (root) and plumule (shoot) in solution. In **hypogeal** germination, e.g. broad bean, the cotyledons remain below the ground. In **epigeal** germination the cotyledons come above the ground (e.g. sunflower). Monocotyledons (e.g. maize) store their food reserves in the endosperm (see figure 110).

Chapter 10. Genetics and evolution

Organisms vary, i.e. differ from one another. This may be continuous, where characteristics show a graduation from one extreme to another, e.g. height in humans; or discontinuous, where characteristics fall into distinct groups, e.g. blood groups in humans. Variation may be inherited or may be due to the environment or both, e.g. rats may be small because of their genes or because they were not receiving enough vitamins in their diet.

Genes are the inherited information carried on the chromosomes. The set of chromosomes in an organism is the genotype, and this reacting with the environment produces the appearance, the phenotype. Every species has a constant number of chromosomes which may be arranged in pairs (diploid). Opposite genes of a pair control the same characteristic, e.g. coat colour; one gene may be for brown coat and the other for fawn coat. One gene will be dominant, e.g. brown coat (B), and will develop to the exclusion of the other recessive gene, e.g. fawn coat (b). Where both genes are of the same type, e.g. for brown coat (BB) or fawn coat (bb) it is homozygous (pure breeding); where not the same (Bb) it is heterozygous. When gametes form the chromosome number is halved so they contain a single set of chromosomes (haploid). (In man diploid is 46, haploid is 23). Separation of genes at cell division (meiosis) and their union at fertilisation is random (by chance). Mutations are changes in genes. The first important breeding experiments were carried out in the nineteenth century by the Austrian monk Gregor Mendel.

Evolution

The **theory of evolution by natural selection** was announced in 1858 by **Charles Darwin** and Wallace. It can be summarised:
1. Members of a species show variation; some can be inherited.
2. Offspring are always more numerous than their parents.
3. Numbers of a species tend to remain more or less constant.
4. There must be a continuous **struggle for existence**.
5. This results in the **survival of the fittest**. Organisms with the most successful variations survive and have offspring which may inherit these variations.

It is only **heritable variations** which are important in evolution and their source was not known to Darwin. Genetic variation arises by mutation (a change in a gene or chromosome) during meiosis at gamete formation, at fertilisation.
Evidence for evolution – fossils, comparative anatomy, etc.
Peppered moth demonstrates natural selection in action; in the last 100 years a dark melanic form has replaced the normal mottled form, the former now having a selective advantage when resting on soot-covered trees (well camouflaged).
Formation of new species – populations become isolated, genes are not exchanged and genetic differences accumulate, and the members may not be able to breed successfully with other populations; this is a stage in the formation of a new species.

Chapter 11. Ecology

This is the study of relationship between organisms and their environment. Different methods of feeding result in organisms within a habitat being linked in food chains and webs. These start with producers (photosynthetic plants) followed by primary consumers (herbivores) followed by secondary consumers (carnivores) and end with decomposers (bacteria and fungi). Energy flows through a habitat, originating from the sun, but chemicals are recycled (illustrated by carbon and nitrogen cycles). **Soil** – consists of inorganic particles, water, air, humus and living organisms (mainly bacteria). Loams contain humus and the correct proportion of small clay particles and larger sand particles so that air and water content are suitable for plant growth. Important to maintain soil fertility, since plants remove mineral elements essential for their growth.

Micro-organisms: bacteria, viruses, fungi

They are microscopic organisms which do not have typical plant or animal cells. They can be classified with the plants although they have heterotrophic nutrition. They may be saprophytes, obtaining food from dead sources, or parasites obtaining food from living organisms (hosts). Those causing disease are known as pathogens. They are measured in micrometers $1\mu m = 1/1000$ mm. **Bacteria** Very small ($0.8 - 20\mu m$), with a cell wall of protein and fat enclosing nuclear material. Some are aerobic, others are anaerobic. In good conditions they can divide into two every twenty to thirty minutes and many can form resistant stages called spores. They are classified by their shape. Cocci (round): in clusters, e.g. staphylococci, causing boils, food poisoning: in chains, e.g. streptococci, causing tonsillitis. Includes bacteria causing pneumonia.
Bacilli (rod-shaped), e.g. typhoid, tuberculosis.
Spirilla (spiral shaped).
Vibrio (comma shaped), e.g. vibrio of cholera.
Some bacteria can be harmful and cause disease and food decay, but many are harmless and others beneficial. They are very important in decay, in sewage disposal and the manufacture of cheese. Bacteria in the gut of herbivores synthesise vitamins and digest cellulose, those in root nodules fix nitrogen. (These are examples of symbiosis – both partners benefit).

Viruses Much smaller than bacteria and only visible with the electron microscope. They consist of a protein coat enclosing nucleic acids. They only reproduce inside living cells, outside they are inert and may take on a crystalline form. Some cause disease, e.g. influenza, colds, measles, poliomyelitis, mumps, chicken-pox. **Fungi** Simple plants lacking cellulose and chlorophyll. Many are saprophytes, others parasites and they reproduce by spores. Very important in decay but some cause disease, e.g. ringworm, athlete's foot, potato blight. The antibiotic penicillin is prepared from the mould *Penicillium*.

Transmission of micro-organisms

They may enter the body through the respiratory passages, the alimentary canal or the skin.

Droplet infection (airborne spread) Occurs when droplets containing micro-organisms are exhaled and then inhaled by other persons. This method transmits the viral and bacterial diseases mentioned. Prevention: good ventilation, personal care, avoidance of crowds.

Food and waterborne spread Contaminated food and water can spread typhoid, cholera, food poisoning and diarrhoea. Food can be contaminated by infected food handlers who exhale over or touch food, by insects (flies) or by infected sewage. Water may be contaminated directly by faeces or urine, or by infected sewage, and used for drinking or food preparation. Food may be inefficiently preserved. Prevention: by personal hygiene, such as washing hands before handling food, and protecting cuts; by keeping food in refrigerators away from flies; by cooking food well and preserving it efficiently; sewage disposal must be efficient, and water supplies purified effectively.

Skin Skin diseases, e.g. ringworm, may be transmitted by direct contact. Cuts may become infected by pus forming staphylococci. Some insects can transmit disease when they bite, e.g. mosquitoes transmit malaria.

Defences of the body against micro-organisms

The skin provides a barrier against micro-organisms. White corpuscles: certain ones, e.g. polymorphs, can move to sites of infection and ingest and kill bacteria; this is called phagocytosis and the cells phagocytes. Certain bacteria produce toxins which are poisonous. Antibodies and antitoxins: are produced in lymph tissue in response to the micro-organisms or their toxins (antigens). Antibodies may kill the bacteria or make them more susceptible to phagocytosis. Antitoxins (a type of antibody) neutralise the poisonous toxins and make them harmless.

Preservation of food

Food is an excellent medium for growth of bacteria and fungi. To preserve food they must be destroyed and further attack prevented, or a condition required for their growth removed (oxygen, warmth, moisture).

Dehydration Water is removed from the food, e.g. milk, eggs, fruit, vegetables. The food is light and easily transported.

Reduced temperature Refrigeration slows and freezing prevents the metabolism of micro-organisms, so slowing or preventing growth. A vast range of food is preserved relatively cheaply, and the food is like the original when thawed. *N.B.* Micro-organisms are not destroyed by dehydration or freezing. Food must be used soon after reconstitution or thawing.

Heating Canning: the principle is to sterilise the food, destroy micro-organisms and seal the cans which cannot be reinfected. Cans keep many years, but incomplete sterilisation can cause food poisoning. Pasteurisation: milk is heated to 72°C for 15 seconds, quickly cooled to 10°C and bottled; this destroys most bacteria.

Addition of preservatives Chemicals kill bacteria and prevent further attack. Only 'permitted preservatives' are used with specific foods, e.g. sausages contain sodium bisulphite. Smoking: Meat or fish is suspended in wood smoke and the creosote kills the bacteria. Salting: Meat or fish immersed in strong salt solution. Modern techniques of radiation and transporting food in air conditioned chambers are being introduced.